Public Policy and the
Diffusion of Technology

The Pennsylvania State University Studies No. 43

Public Policy and the Diffusion of Technology

An International Comparison of Large Fossil-Fueled Generating Units

by John H. DeYoung, Jr.
and John E. Tilton

The Pennsylvania State University Press
University Park and London

Library of Congress Cataloging in Publication Data

DeYoung, John H., Jr.

Public policy and the diffusion of technology.

(The Pennsylvania State University studies; no. 43)

Based on the thesis of J. H. DeYoung, Jr., Pennsylvania State University.

Includes bibliography.

1. Energy policy. 2. Electric utilities.

I. Tilton, John E., joint author. II. Title.

III. Series: Pennsylvania. State University. The Pennsylvania State University studies; no. 43.

HD9502.A2D477 338.4'7'62131 78–50067

ISBN 0–271–00547–5

Printed in the United States of America

Contents

Preface

Over the last few years energy has emerged as one of the most perplexing and pressing problems confronting governments around the world. Not surprisingly, this new concern has stimulated a plethora of studies on the future availability of energy.

Much of the recent research is vast in scope. This study, in contrast, focuses on a specific phenomenon—the adoption of new, energy-saving generating units in the electric power industry. It was our hope in undertaking this investigation that a detailed case study of one particular type of new technology in one particular industry might provide insights into the process by which new technology, especially new energy-saving technology, is diffused, and that these insights in turn might facilitate the formulation of public policies to accelerate the adoption of such technology.

This study draws heavily on a doctoral dissertation prepared by John H. DeYoung, Jr., for the Department of Mineral Economics at The Pennsylvania State University, and the research he conducted for that dissertation. It, along with Mr. DeYoung's dissertation, was funded by the Center for Policy Alternatives at the Massachusetts Institute of Technology. We are grateful to J. Herbert Hollomon, Director of the Center for Policy Alternatives, for both his financial support and his assistance in defining the project during a conference at MIT in 1974.

Many people generously made their time available during interviews and offered valuable information and encouragement. The organizations to which we are especially indebted include the U.S. Federal Power Commission, General Electric Company, Westinghouse Electric Corporation, Atlantic City Electric Company, West Penn Power Company, and the Pennsylvania-New Jersey-Maryland Interconnection, all in the United States; the Department of Energy, Mines, and Resources, the Department of Industry, Trade, and Commerce, and the Hydroelectric Power Commission of Ontario, all in Canada; and the Central Electricity Generating Board, the Electricity Council, and the Science Policy Research Unit of the University of Sussex, all in Great Britain.

We are also grateful to Terry Boley, Joel Darmstadter, and DeVerle P.

Harris for their comments on earlier drafts of this study, and to colleagues at The Pennsylvania State University, especially Irwin Feller, Richard L. Gordon, John D. Ridge, and William A. Vogely, for their helpful suggestions. Sally Jewett DeYoung, Karen Flynn, and Barbara Poorman carefully typed various versions of the manuscript.

In acknowledging the valuable assistance we have received from many sources, we wish to implicate no one. The views expressed in this study are our own and do not necessarily represent those of the individuals or organizations that have assisted or supported us.

1
Introduction

The sharp rise in the cost of energy over the last few years coupled with the fear of another Arab oil embargo have greatly increased the interest of industrialized countries in accelerating the development of new energy technology. Since the early 1970s, for example, the United States government has more than tripled the amount of money its spends annually on energy research and development in order to reduce its dependence on foreign sources of energy supply (Tilton, 1976).

Although it has received far less attention, another important activity that could help relieve energy needs involves the more rapid and widespread use or diffusion of existing energy-saving technology. As this technology is already developed and ready for commercial use, its benefits can be realized without the long delays typically associated with the creation of new technology.

This book investigates the adoption of new technology in the electric utility industry. Specifically, it identifies and assesses the relative importance of the major factors influencing the rate at which large-scale generating units have been introduced. It also considers how public policy might influence these factors so as to stimulate the rate of adoption.

The electric power industry was chosen for several reasons. First, it is important and growing more important with time. Since World War II, rapid growth in the demand for electric power has allowed the industry to double its output every ten years, and it now ranks among the largest industries in terms of both sales and value of assets. Second, governments have always played a major role in the supply of electricity both by regulation of privately owned firms and by ownership of power systems. Consequently, public policy greatly affects the behavior of this industry. Third, the interregional and international differences that exist in this industry provide an opportunity to study the diffusion of new technology under different industrial structures, types of ownership, demand patterns, and supply conditions. Finally, information about the equipment installed by electric utilities is readily available from government agencies, trade journals, and the utilities themselves.

The scope of this book is limited in several ways. First, no attempt is made to explain the creation of new generating technology. The focus is entirely on diffusion. Second, the analysis concentrates on the adoption of one innovation, or set of innovations: large fossil-fueled steam-electric generating units. Since increases in the pressures and temperatures at which these units operate have also increased with unit size and account at least in part for the resulting increases in efficiency, they too are considered. However, other technological developments, such as reheat stages and improved firing methods, are not examined. Third, the book investigates diffusion in three countries—the United States, Canada, and Great Britain. In the first two countries, differences in diffusion among geographic regions are considered as well. Fourth, the analysis focuses on the adoption of generating technology over the last quarter century.

Chapter 2 provides background information on the electric power industry needed in later chapters. Chapter 3 then considers the nature of the diffusion process. Hypotheses are developed in this chapter concerning the important factors influencing the diffusion of new generating technology.

Chapters 4, 5, and 6 examine the adoption of new generating units in the United States, Canada, and Great Britain, respectively. These chapters consider the extent to which the hypotheses or expectations advanced in Chapter 3 are, in fact, realized in these countries. They also identify several important factors influencing the rate of adoption that the analysis of Chapter 3 fails to anticipate.

Chapter 7, the final chapter, summarizes the major findings. In addition, it examines the potential for public policy to stimulate the adoption of new energy-saving technology in the electric power industry.

2
The Electric Power Industry

After identifying the stages of electric power production, this chapter describes the principal methods of generating electricity. It then examines technological advances in the steam generation of electric power from fossil fuels, with particular emphasis on the development of larger generating units since 1950. Finally, it briefly investigates the organization of the electric power industries in the United States, Canada, and Great Britain, and concludes with several observations of particular importance for this study on the electrical equipment industry.

Stages of Production

There are three stages in the production of electricity: generation, transmission, and distribution. The first creates electric power; the second transports it over bulk power lines at high voltages; and the third distributes it at low voltages to local consumers.

Most of the costs of supplying electric power are incurred in generation. Table 1 shows that this stage of production accounted for over half of the costs of producing electricity in the United States in 1968. Since then, higher fuel costs have increased the share of costs incurred at this stage.

The production of electric power takes place in a generating station where the mechanical force of high-pressure steam or moving water drives a turbine. The turbine is connected to a generator that creates electrical energy from the kinetic energy of the rotating shaft. The turbine and generator are often installed as a single unit referred to as a turbine-generator set.

When steam is used to power turbine-generators, it is usually produced in boilers heated by coal, oil, natural gas, wood, peat, industrial waste, or nuclear fission, though in a few areas geothermal steam is also available.

Table 1

Costs of Producing Electricity in the United States, 1968

	Total Cost (millions of dollars)	Unit Cost (mills/kwh)	Percentage of Total
Generation			
Fuel	2,960	2.47	
Other operation and maintenance	1,609	1.34	
Allocated administration and general	278	0.23	
Fixed charges	4,443	3.71	
Total generation cost	9,290	7.75	50.3
Transmission			
Operation and maintenance	295	0.25	
Allocated administration and general	86	0.07	
Fixed charges	1,995	1.66	
Total transmission cost	2,376	1.98	12.8
Distribution			
Operation and maintenance	1,963	1.64	
Allocated administration and general	585	0.49	
Fixed charges	4,270	3.56	
Total distribution cost	6,818	5.69	36.9
Total power cost	18,484	15.42	100.0

Source: U.S. Federal Power Commission, 1971, p. I–19–10.

In addition to steam and hydropower, internal combustion engines and gas turbines are used on a small scale to produce electricity.

Despite the growing importance of nuclear power over the last decade and the abundance of hydropower in certain regions, fossil fuels still produce about three-quarters of the world's electric power. In the United States, as Table 2 shows, fossil-fueled steam capacity accounted for 75% of the electric power produced in 1972; in Great Britain, the figure was 90%. One exception of particular importance for this study is Canada

Table 2
Fossil-Fueled Steam Generation of Electric Energy as a Percentage of Total Generation in the United States, Canada, and Great Britain, 1972

	Electric Energy Production (Twh)[a]		Percentage of
Country	Total	Fossil-Fueled Steam	Generation from Fossil-Fueled Steam
United States	1,853.4	1,378.3	74.4
Canada	240.2	52.1	21.7
Great Britain[b]	204.5	183.3	89.6

[a]Terawatt-hours; 1 terawatt-hour = 1 billion kilowatt-hours.

[b]Central Electricity Generating Board production for year ended 31 March 1973.

Source: Edison Electric Institute, 1974; U.S. Federal Power Commission, *Steam-Electric Plant Construction Cost and Annual Production Expenses*, 1972; Statistics Canada, 1974b; Central Electricity Generating Board, 1973a.

where hydropower produced 75% of the country's electric power and fossil-fueled steam capacity only 22%.

Over time, however, the relative importance of hydropower has been declining in Canada and elsewhere as the number of undeveloped hydropower sites diminishes. This along with the slower than expected growth in nuclear power and the new interest in coal, particularly in the United States, suggests that fossil-fueled steam generation will remain a major source of electric power for the foreseeable future.

Advances in Fossil-Fueled Generation Technology

Over time technological progress has greatly improved the efficiency of electric power generation. One of the most important developments in this regard has been the introduction of larger generating units. In 1903, when Commonwealth Edison equipped its Fisk station, the first utility station designed exclusively for turbine generators, it installed four 5-MW units. Seventy years later, the Tennessee Valley Authority began operating a

Table 3

Size of Largest Generating Unit, Advanced Practice Scale Class, and Established Scale Class in the United States, 1950–76

Year	Largest Pioneering Unit to Date (MW)[a] $N \geq 1$[b]	Largest Advanced Practice Scale Class (MW)[c] $N \geq 7$	Largest Established Scale Class (MW)[c] $N \geq 15$
1950	165	100	80
1951	165	125	80
1952	165	125	125
1953	180	150	125
1954	180	175	125
1955	207	175	175
1956	260	175	175
1957	260	200	175
1958	335	325	250
1959	335	325	250
1960	450	325	250
1961	575	400	325
1962	575	400	325
1963	704	400	400
1964	704	400	400
1965	1,027	400	400
1966	1,027	580	400
1967	1,027	580	580
1968	1,027	650	580
1969	1,027	650	580
1970	1,150	815	650
1971	1,150	815	815
1972	1,300	815	815
1973	1,300	950	815
1974	1,300	1,100	815
1975	1,300	1,100	815
1976	1,300	1,100	815

[a] Nameplate capacity.

[b] N represents the number of units in service that belonged to the listed scale class.

[c] Scale classes include the following ranges:

Scale class (MW)	Range (MW)	Scale class (MW)	Range (MW)
80	75–85	325	290–335
100	100–110	400	359–450
125	112.5–135	470	454.8–517.5
150	145–156	580	523.8–615.2
165	165	650	620.5–693.9
175	169–180	750	700–765
200	200–225	815	781–850
250	231.3–275	950	892.8–952
		1,100	1,000–1,300

Source: Hughes, 1970, p. 163; U.S. Federal Power Commission, *Steam-Electric Plant Construction Cost and Annual Production Expenses.*

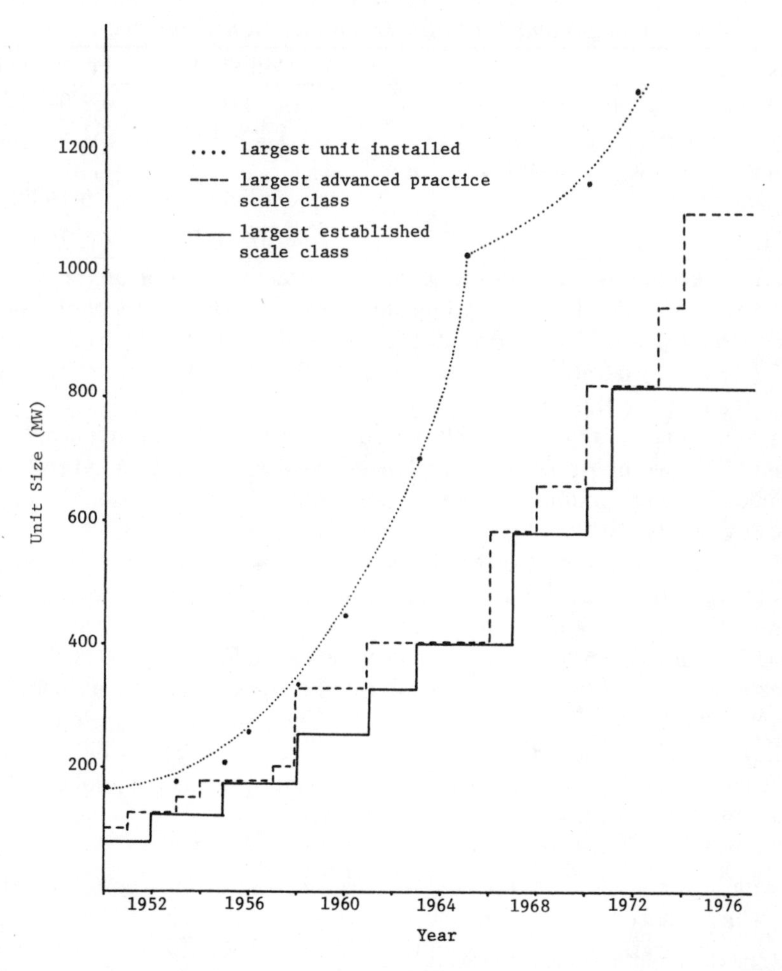

Figure 1. Size of the largest generating unit, advanced practice scale class, and established scale class in the United States, 1950–76. (*Source:* Table 3.)

1300-MW unit at its Cumberland station, and other such units were under construction.[1]

For a number of years since 1950, Hughes (1970, p. 163) has identified for the United States (1) the largest generating unit in operation, (2) the largest advanced practice scale class containing seven or more operating units, and (3) the largest established scale class containing 15 or more

Table 4
Component Costs for Electricity Generation[a]

Component	Mills/kwh	Percentage
Annualized capital outlays[b]	4.38	62.0–44.9
Fuel expenses[c]	2.25–4.95	31.9–50.7
Operation and maintenance expenses[b]	0.43	6.1–4.4
Total	7.06–9.76	100

[a] More recent cost estimates for coal or oil-fired production are provided by the U.S. Federal Energy Administration's 1974 *Project Independence Blueprint* (p. VII–72). Total production costs of 23.21 mills/kwh (coal) and 27.09 mills/kwh (oil) that are predicted for 1980 are nearly three times the figures given in this table.

[b] From Hughes (1970, p. 99) based on figures for a hypothetical 400-MW, two-unit coal burning plant of 1964 vintage presented in U.S. Federal Power Commission, 1964, Table 27, p. 70. Annual costs of capital outlays are based on a "fixed charge rate" of 13%. Rates of 12–14% have been commonly used by privately owned utilities to cover cost of capital, depreciation, and capital-related taxes (see U.S. Federal Power Commission, 1971, p. I–19–6). Assumed annual capacity factor is 65%.

[c] The assumed fuel prices are 25 and 55 cents/million Btu, which covers the range of 1972 regional average fossil fuel costs for electricity generation in the United States (see Chapter 4). The assumed heat rate is 9000 Btu/kwh. Using the national average fuel price of 39.9 cents/million Btu, the capital, fuel, and operation and maintenance portions of total generation costs are 52.2, 42.7, and 5.1%, respectively.

operating units. His data updated to 1976 are presented in Table 3. Figure 1, which is based on this table, illustrates that over the last quarter century the available size of generating units has increased continually and that the increases have been substantial.

Larger units have been developed and introduced primarily because they reduce both the capital and fuel needed to produce electricity. As seen in Table 4, these two important inputs account for over 90% of generation costs. The third component of generating costs, maintenance and operating costs other than expenditures on fuel, also tends to fall with bigger units, though it amounts to only a small fraction of generating cost.

The U.S. Federal Power Commission (1971, p. I–19–4) has estimated that capital costs in terms of 1968 price levels decline from $210 to $160/kw of installed capacity, or by nearly 25%, as the size of fossil-fuel

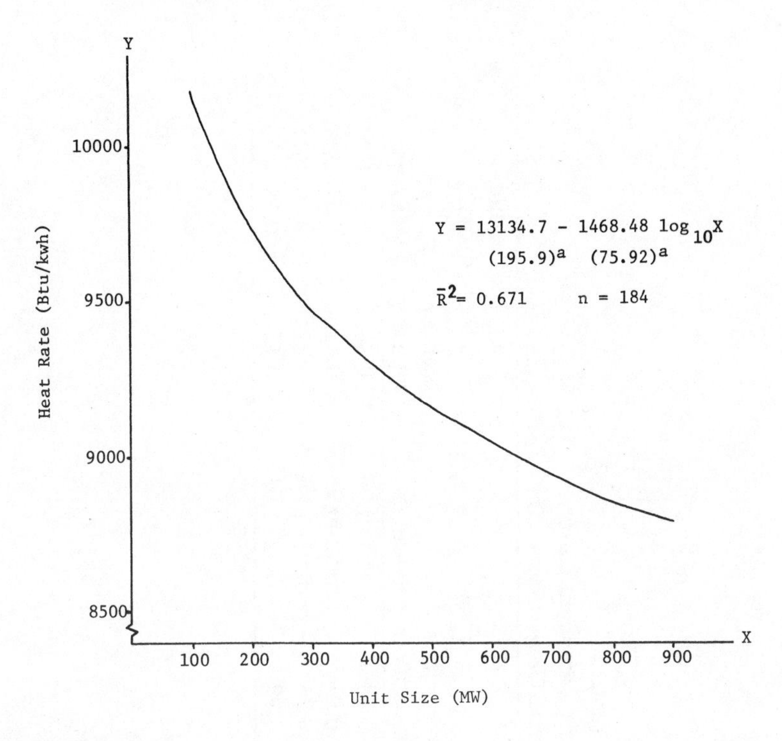

[a] Standard errors.

Figure 2. Relationship of heat rate and rated capacity for fossil-fueled steam-electric generating units installed by utility systems in the United States, 1968–72. (*Source:* Calculated on the basis of data from General Electric EUPB data bank.)

steam units increases from under 100 to over 900 MW. Although inflation and environmental costs have increased the absolute costs of new capacity since these estimates were made, the proportional decrease in costs associated with larger unit size is still substantial.

Savings in fuel are reflected by the heat rates of generating units, which indicate the number of British thermal units of fuel required to produce a kilowatt-hour of electricity (Btu/kwh). To estimate the relationship between unit size and heat rates, the heat rates of 184 units installed in the United States between 1968 and 1972 were regressed on the rated capacities of those units. The log-linear specification of this relationship is presented in Figure 2. Its corrected coefficient of determination $\bar{R}^2$ suggests that unit size can account for 67% of the variation in heat rates

Table 5
Elasticities of Component Costs for Steam-Electric Generating Units

Scale Range	Capital: Average Elasticity	Capital: Range	Fuel: Average Elasticity	Fuel: Range	Operation and Maintenance: Average Elasticity	Operation and Maintenance: Range
50–100	0.74	0.65–0.83	0.88	0.85–0.90	0.48	0.30–0.51
100–200	0.78	0.75–0.84	0.93	0.92–0.96	0.44	0.31–0.52
200–400	0.78	0.64–0.88	0.96	0.93–1.00	0.42	0.16–0.58
400–800	0.87	0.81–0.94	0.99	0.92–1.02	0.52	0.18–0.73

Source: Hughes, 1970, p. 104.

among the units analyzed. Linear and quadratic specifications (with $\bar{R}^2$s of 0.35 and 0.46, respectively) did not do as well. The log-linear relationship between heat rate and unit size implies that larger gains in thermal efficiency are made in advancing from 100- to 300-MW units than from 700- to 900-MW units.

Other evidence concerning the effect of unit size on generating costs has been collected by Hughes (1970, p. 104). He reviewed eleven engineering and econometric studies of cost functions for steam-electric generating units and on the basis of these studies estimated average elasticities for capital, fuel, and operating and maintenance expenses for different scale ranges. These elasticities, given in Table 5, indicate the percentage increase in costs incurred when unit size and output are expanded by 1%. According to these elasticities, an increase in unit size and output results in a less than proportional increase in total costs. This means per unit costs fall as unit size increases. This holds as well for the three components of costs—capital, fuel, and operation and maintenance.

Table 5 also shows that the reduction in unit costs caused by an increase in unit size is greater for smaller units than for larger units. For example, an increase in unit size and output of 10% requires an increase in capital of 7.4%, in fuel of 8.8%, and operation and maintenance expenses of 4.8% in the 50–100-MW scale range. In the 400–800-MW range, however, such an increase requires an increase in capital of 8.7%, in fuel of 9.9%, and in operation and maintenance costs of 5.2%.

While larger units reduce generating costs, they make it more difficult or costly to maintain reliability of service. Since electricity cannot be stored except in small quantities, utilities must install enough reserve capacity to meet demand even when some units are not in service due to scheduled maintenance or unscheduled breakdowns (forced outages). How much reserve capacity a system needs depends on the reliability of its generating units. Reliability is measured by a unit's forced outage rate (FOR), which indicates the amount of time a unit is not in service due to forced outages as a percentage of its service plus forced outage time.

Table 6 shows the increase in FOR's associated with larger unit sizes, as estimated by the U.S. Federal Power Commission and the Edison Electric Institute. Although the latter's figures are appreciably higher than those of the Federal Power Commission, both sets of estimates show that the FOR increases with unit size. Larger units have higher FOR's both because they break down more frequently and because they take longer to repair. The effect of large units on the reliability of service can be severe, particularly for small systems, not only because of greater frequency and

Table 6

Equivalent Forced Outage Rates, Fossil-Fueled Steam-Electric Generating Units in the United States

	Forced Outage Rates[a]	
Unit Size (MW)	EEI[b]	FPC[c]
60–89	1.96	—[d]
90–129	3.82	2.0–2.6
130–199	4.14	2.1–3.0
200–389	6.52	2.3–3.5
390–599	11.33	2.5–4.4
≥600	21.90	3.1–7.0

[a] Forced outage rates (FOR's) indicate the amount of time units are not in service due to forced outages as a percentage of service plus forced outage time (Billinton et al., 1973, p. 84).

[b] Equivalent forced outage rate for units in the United States for 1960–72; data collected by the Edison Electric Institute.

[c] From graph of forced outage rates versus unit size (mature units) prepared by the U.S. Federal Power Commission.

[d] No information on this size range in FPC graph.

Source: Edison Electric Institute, 1973; U.S. Federal Power Commission, 1971, p. IV–1–14.

duration of outages, but also because their removal from service takes a larger proportion of capacity out of the system.

Advances in fossil-fueled steam-electric generating equipment have not been restricted to unit size. Steam conditions, particularly inlet steam pressures and temperatures, have also improved efficiency. Indeed, since increases in steam pressures and temperatures are often associated with larger unit size, some of the savings in capital, fuel, and operating and maintenance costs attributed to larger unit size are actually the result of improvements in steam pressures and temperatures. However, Table 7 shows that steam conditions, especially inlet steam pressure, are highly correlated with unit size. Consequently, unit size can be used as a surrogate or proxy for technological advances in both unit size and steam conditions.

Of course, unit size cannot serve as a surrogate for all improvements in fossil-fueled generating technology, as many advances have not been associated or correlated with increases in unit size. Outdoor production, automation of operations, better generator rotor design, tilted tangential firing, reheat and double reheat cycles, and controlled circulation are but a

Table 7
Relationship between Rated Capacity, Inlet Steam Pressure, and Inlet Steam Temperature of Fossil-Fueled Steam-Electric Generating Units

	Correlation Coefficients		
	United States	Canada	England[a]
Size-pressure	.844	.863	.868
Size-temperature	.389	.584	.727
Temperature-pressure	.660	.801	.930
Number of units in sample	1,225	150	425
Years units were introduced	1953–72	1949–73	1949–73

[a] Central Electricity Generating Board.

Source: U.S. Federal Power Commission, *Steam-Electric Plant Construction Costs and Annual Production Expenses*, 1953–1972; Statistics Canada, 1974c; Central Electricity Generating Board, 1973b.

few examples (*Electrical World*, 1974, pp. 78–79). Still, advances in the size of generating units along with associated improvements in steam pressures and temperatures constitute one of the most important sets of developments in generation technology over the last several decades. As a consequence, the speed at which larger units have been introduced provides an important indication of the performance of firms in adopting new technology in the electric power industry.

Organization of the Electric Utility Industry

The present organization of the electric power industries in the United States, Canada, and Great Britain is the result of years of development influenced by national and regional resource patterns, the performance of the industry in its formative years, and public policy toward the industry.

United States

The important changes in the structure of the electric utility industry of the United States can be divided into four stages—the formation, consolidation, coordination, and further consolidation of electric power systems (Hughes, 1970). From 1882 to 1900, the formation of local systems was encouraged by inventors and manufacturers of electrical equipment to increase the use of their products. Many of the small companies established during this period were consolidated between 1900 and 1935 by means of the holding company. However, the abuses uncovered by the Federal Trade Commission investigation into the operations of the holding companies in the late 1920s led to the passage of the Public Utility Holding Company Act of 1935. This act provided for the abolition of all holding companies more than twice removed from operating subsidiaries, and arrested the trend toward greater concentration in the industry.

By the 1950s, the industry had been reorganized. It was beginning to pursue the benefits of larger size through greater interconnection and coördination among systems. These activities reduce the need for reserve capacity by allowing systems whose peak loads occur at different times to exchange power. Similarly, the risks of unit failure can be shared. They also lower costs by allowing greater flexibility in the planning and construction of generating capacity. New facilities can be situated where costs and construction delays are minimized. Firms have coordinated and interconnected their systems through pooling agreements, joint generating projects, agreements for mutual assistance, and regional reliability groupings. Pooling agreements run the gamut from simple emergency assistance pacts to complete coordination of all generating units owned by parties to the agreement in order to minimize combined production costs.

While the industry has since World War II pursued the benefits of larger size for reliability and efficiency primarily through more intersystem coordination, there has been a reduction in the number of operating firms and thus some consolidation of the industry. Nevertheless, the number of firms producing electric power remains large. In 1975, the country possessed 505,080 MW of generating capacity, of which 78.0% was owned by the 185 private utilities that operate generating facilities. These firms ranged in size from the Commonwealth Edison Company in Illinois with 16,417 MW of capacity to firms with less than 2 MW of capacity. In addition, electric power is generated by 5 federal agencies (the Tennessee Valley Authority and hydroelectric facilities of the U.S. Corps of Engineers and the Bureau of Reclamation account for 99.8% of federally owned capacity), by about 190 state and locally owned companies, and by about 65 cooperatives. In 1975, these three groups operated 9.5, 8.8, and 3.7% of the country's generating cacacity, respectively.

Canada

The development of the electric transformer in 1897 paved the way for long-distance transmission of hydroelectric power in Canada. Until that time, most electricity was generated in fossil-fueled plants situated close to the points of consumption.

In 1906, the Hydroelectric Power Commission of Ontario was formed when manufacturers in southern Ontario decided that publicly owned hydroelectric plants could distribute and sell power more cheaply than privately owned utilities. Soon after World War I, provincial agencies were established to produce and distribute electric power in New Brunswick, Manitoba, and Nova Scotia. The Commission Hydroelectrique de Québec (Quebec Hydro) and the British Columbia Hydro and Power Authority were formed in 1943 and 1961, respectively, to acquire privately owned utilities in those provinces. In contrast, private, investor-owned utilities still dominate the electric power industries in Alberta, Prince Edward Island, and Newfoundland, though the recent sale of the rights to develop the Churchill River basin in Labrador by an investor-owned utility to the provincially owned Newfoundland and Labrador Power Commission may portend an increasing role for publicly owned utilities there. In any case, as Table 8 illustrates, most of Canada's generating capacity is concentrated in Ontario, Quebec, and British Columbia, where provincial ownership predominates. As a result, over 75% of Canada's generating capacity is controlled by publicly owned firms.

The Canadian electric power industry differs from its counterpart in the United States in another respect, the importance of hydroelectric power. Ninety-two percent of Canadian generating capacity in 1950 was hydroelectric. By 1973, as shown in Table 8, its share had fallen, though it still accounted for 63% of capacity; the remainder was steam-electric (34%) and internal combustion and gas turbine (3%). Only Ontario, Alberta, Saskatchewan, Nova Scotia, and Prince Edward Island have more steam than hydroelectric capacity. Moreover, hydroelectric facilities account for an even larger share of electric power production since they are used primarily as base-load units and so operated on a more continuous basis than steam-electric generating units.

Great Britain

Private ownership of electric utilities in Great Britain was discouraged by the first Electric Lighting Act in 1882. A public corporation was estab-

Table 8

Generating Capacity in Canada by Province, Type of Ownership, and Nature of Equipment as of 31 December 1973 (MW)

	Ownership			
	Public	Private	Industry	Total[a]
All Equipment[b]				
Newfoundland	825.0	3,834.8	84.1	4,743.9
Prince Edward Island	6.9	111.4	—	118.2
Nova Scotia	1,112.9	—	90.6	1,203.4
New Brunswick	1,175.7	31.8	124.2	1,331.7
Quebec	11,447.0	680.4	2,711.9	14,839.3
Ontario	16,674.9	340.0	792.2	17,807.2
Manitoba	2,657.3	—	7.0	2,664.3
Saskatchewan	1,623.4	106.7	44.8	1,774.9
Alberta	837.0	2,412.9	154.1	3,404.0
British Columbia	4,453.6	48.7	1,811.8	6,314.2
Northwest Territories	98.6	6.6	7.4	112.6
Yukon	47.0	8.1	7.5	62.6
Total[a]	40,959.3	7,581.5	5,835.6	54,376.4
Hydroelectric				
Newfoundland	459.9	3,775.8	64.1	4,229.8
Prince Edward Island	—	—	—	—
Nova Scotia	155.3	—	5.0	160.3
New Brunswick	634.8	30.8	14.2	679.9

Quebec	10,496.2	680.4	2,622.3	13,798.9
Ontario	6,442.8	331.7	253.5	7,007.9
Manitoba	2,169.1	—	—	2,169.1
Saskatchewan	447.8	106.7	12.3	566.9
Alberta	—	718.3	—	718.3
British Columbia	3,326.6	48.5	1,428.1	4,803.2
Northwest Territories	32.0	—	3.4	35.4
Yukon	24.5	1.6	—	26.1
Total[a]	24,169.1	5,693.9	4,402.9	34,265.9
Steam				
Newfoundland	308.0	30.0	16.6	354.6
Prince Edward Island	—	70.5	—	70.5
Nova Scotia	926.8	—	85.0	1,011.7
New Brunswick	510.6	—	110.0	620.6
Quebec	866.0	—	76.2	942.2
Ontario	9,864.2	—	427.8	10,292.0
Manitoba	443.0	—	4.0	447.0
Saskatchewan	1,063.0	—	23.0	1,086.0
Alberta	749.3	1,567.0	130.2	2,446.5
British Columbia	810.0	—	339.7	1,149.7
Northwest Territories	0.6	—	—	0.6
Yukon	—	—	—	—
Total[a]	15,541.5	1,667.5	1,212.3	18,421.3

[a] Data may not add to totals shown due to rounding.

[b] Includes hydroelectric, steam, internal combustion, and gas turbine.

Source: Statistics Canada, 1974c, p. 15.

Table 9

Generating Capacity in Great Britain Operated by the Central Electricity Generating Board by Regions and Type of Equipment as of 31 March 1974

	Regions Classified by Type of Station						
	Steam (Conventional) MWI[a]	Steam (Nuclear) MWI	Diesel MWI	Gas Turbine MWI	Hydro and Pumped Storage MWI	Regional Total[b] Number of Stations	Regional Total[b] MWI
Northwestern	6,044	1,586		183	470	34	8,284
Northeastern	11,278			336		31	11,614
Midlands	16,378			340		35	16,718
Southeastern	11,444	1,603	10	640		41	13,698
Southwestern	10,178	1,632		436	6	28	12,251
Total CEGB	55,321	4,821	10	1,936	476	169	62,564

[a] MWI = megawatts installed, which is equivalent to nameplate capacity.

[b] Data may not add to totals shown due to rounding.

Source: Central Electricity Generating Board, *CEGB Statistical Yearbook*, 1973–74, p. 6.

lished to construct a national power grid in 1926, and in 1947 electricity supply in England, Wales, and southern Scotland was nationalized. The Electricity Act of 1957 placed responsibility for the generation and main transmission of electric power as well as for the coordination and policy direction of the industry with two statutory bodies, the Electricity Council and the Central Electricity Generating Board.

The Electricity Council sets overall policy for the industry. Its duties are to advise the Secretary of State for Trade and Industry concerning electricity supply and "to promote and assist the maintenance and development by Electricity Boards in England and Wales of an efficient, coordinated, and economical system of electricity supply." Its areas of responsibilities include forecasting demand for future investment requirements, reviewing proposed changes in electricity rates, examining industrial relations, and conducting research.

The Central Electricity Generating Board (CEGB) owns and operates the generating and main transmission facilities. In early 1974, it had 62,564 MW of capacity located in 169 power stations. For operation and administrative purposes the CEGB is divided into five geographic regions. Table 9 shows the amount and type of capacity for each of these regions.

The electric power industry of Great Britain is similar to that of the United States with respect to its heavy reliance on fossil-fueled steam-generating capacity, and it is similar to that of Canada with respect to the predominance of public ownership. At the same time, it differs from both of the North American industries in that it is completely controlled by one firm.

Producers of Electrical Equipment

Electric utilities in the United States, Canada, Great Britain, and other countries purchase equipment for the generation, transmission, and distribution of electricity from independent manufacturers of electrical equipment. Table 10 identifies the producers of heavy electrical generators outside of the centrally planned countries. These firms compete for business around the world.

The fact that utilities buy their equipment from independent producers rather than produce it themselves is important, for if this were not the case some utilities might be slow to introduce large generating units because their in-house equipment divisions could be late in developing such units. Similarly, the willingness of equipment producers to install new units for any utility regardless of whether it is located in the United States, Canada,

Table 10
Manufacturers of Heavy Electrical Turbine Generators, 1972

Country	Company
United States	General Electric Company
	Westinghouse Electric Corporation
Canada	Canadian General Electric Ltd. (CGE)[a]
	Howden-Parsons Ltd.[b]
United Kingdom	Reyrolle-Parsons
	General Electric Company Ltd. (GEC)[c]
France	Rateau-Schneider
	Alsthom
	Compagnie Electro-Mécanique[d]
Italy	Franco Tosi
	Ercole Marelli
	Ansaldo
	Tecnomasio Italiano Brown Boveri[d]
	Breda
West Germany	Brown Boveri[d]
	Kraftwerk Union
	Maschinenfabrik Augsberg-Nurnberg AG
Sweden	Stal-Laval[e]
Switzerland	Brown Boveri
Japan	Hitachi
	Toshiba
	Mitsubishi

[a] Subsidiary of General Electric Company (United States).
[b] Subsidiary of Reyrolle-Parsons (United Kingdom).
[c] English Electric merged into GEC in 1968.
[d] Subsidiary of Brown Boveri (Switzerland).
[e] Subsidiary of Allmänna Svenska Elektriska Atiebolaget (ASEA) (Sweden).
Source: Epstein, 1972, p. 39.

or Great Britain means that utilities in these countries have access to the latest and largest generating units available. Thus accessibility, or the lack of it, should not inhibit the adoption of large generating units in any of these countries.

3

Incentives to Adopt New Generating Technology

In recent years, some economists have argued that in making decisions, managers, regardless of the nature of their firm or organization, attempt to maximize their own welfare or utility functions.[1] These functions depend on salary, other forms of pecuniary compensation, job security, prestige, power, freedom from outside interference, and possibly a sense of contributing to the social weal. In the electric power industry, attaining these objectives depends primarily on the cost and the reliability of service. Interruptions in service create public outcries, increase the likelihood of government investigations, and threaten job tenure. Similarly, rising costs mean higher prices, consumer opposition, and greater public surveillance. On the other hand, reliable service at declining costs is likely to enhance managerial remuneration, prestige, and independence.

Thus in considering what type of new generating equipment to install, managers are greatly influenced by the expected effect of the available alternatives on generating costs and reliability. This suggests that international and interregional differences in the rate at which large and technically advanced generating units are introduced arise largely because the effect of these units on the cost and reliability of electric service differs between countries and even regions within a country.

The reasons for this are examined in this chapter. The first section looks at generating costs; the second at system reliability.

Savings in Generation Costs

Chapter 2 pointed out that the generation stage historically accounted for over half the cost of electricity (Table 1), and that the major costs at this stage of production were for capital (52%) and fuel (43%). Operation and maintenance expenses, including labor costs, are responsible for only 5% of generating costs (Table 4). Increases in the size of steam-electric

Table 11

Comparison of Production Costs for 420-MW and 630-MW Units ca. 1970

	420-MW Unit		630-MW Unit[b]		Differences	
	Cost ($/kw)	% of Total[a]	Cost ($/kw)	% of Total	Δ Cost ($/kw)	% of Reduction
Capital	190[c]	52.2	181.8	51.6	8.2	4.3
Fuel	155.4	42.7	154.9	44.0	0.5	0.3
Operation and maintenance	18.6	5.1	15.6	4.4	3.0	16.1
Total	364.0	100.0	352.8	100.0	11.7	3.2

[a] From Table 4.

[b] Costs calculated using elasticities for the 400–800-MW range from Hughes (1970, p. 104); see Table 5.

[c] Estimated from U.S. Federal Power Commission, 1971, p. I–19–4.

generating sets reduced all three components of costs. While the percentage of reduction in operation and maintenance costs exceeds that of either capital or fuel, greater absolute amounts are saved on the latter since these two factors together account for 95% of generating costs.

A system adding 1260 MW of capacity in a year can choose between three 420-MW units or two 630-MW sets. Assuming a cost of $190/kw of installed capacity as the capital cost of the 420-MW unit, as the U.S. Federal Power Commission has done (U.S. Federal Power Commission, 1971, p. I–19–4), fuel and operating and maintenance costs can be estimated using the composition of generating costs presented in Chapter 2. The results are shown in Table 11.

This table also shows the estimated costs associated with a 630-MW unit. This larger unit size realizes savings in capital costs of 4.3%, fuel costs of 0.3%, operation and maintenance costs of 16.1%, and total costs of 3.2% over units of the 420-MW class. Moreover, the savings in fuel are based on 1972 data and do not reflect the rapid rise of fuel costs since then. For this and probably other reasons, they are underestimated. When the regression equation in Figure 2 is used to estimate the fuel saved by using 630-MW units instead of 420-MW units, a 2.8% reduction is obtained.

This analysis assumes that the savings associated with individual generating units provide the incentive for the firm to adopt larger, more efficient generating units. Under some circumstances it is conceivable that total savings associated with the introduction of all units of a given size would be the major consideration. This could occur when substantial learning costs are associated with the adoption of a particular size unit. Larger firms would have a larger incentive to introduce more of the technologically advanced units, since they could use more of these units and thus spread any transitional or learning costs over a greater number of units than could a smaller firm.

However, of the fossil-fueled steam-electric generating units installed in the United States from 1953 through 1972, less than 9% were larger than the largest advanced practice scale class as defined by Hughes (see Figure 1). Consequently, for most units introduced there were units of similar sizes already in operation for which information was readily available. In addition, larger firms often are not able to spread the learning costs associated with a given scale over a large number of units because the rapid advance of technology in this industry results in the introduction of an even larger scale class soon after a new scale class has been introduced. For these reasons, this study assumes that the cost savings which stimulate adoption are those associated with individual generating units.

The nature of generating costs and the savings derived from larger units suggest that interregional and international differences in the cost of capital (interest rates) and the cost of fossil fuel should be the major cost considerations effecting differences in the speed with which large generating units are adopted. The costs of fossil fuels differ considerably among countries and regions. In 1972, the British Central Electricity Generating Board paid an average of almost 65 U.S. cents per million Btu, 294% higher than the 16.5 cents paid by utilities in western Canada. Regional fuel prices in the United States in 1972 ranged from 55 cents in New England to 25 cents in the West South Central states. As a result of such large differences plus the importance of fuel costs in generating electricity and the fuel-saving ability of larger units, one would expect fuel costs to be an important determinant of adoption behavior. Utilities in regions or countries with high fuel costs should adopt large units more rapidly than those in areas with low fuel costs.

International differences in the cost of capital to utilities are assumed to be small compared to those for fuel costs. This is in part because there are few barriers to the movement of capital, particularly between advanced countries such as those examined in this study, and in part because the cost of moving capital is negligible. Money market and Eurodollar rates over the past decade have been higher in the United Kingdom than in Canada and the United States. Rates in the United States and Canada have been similar and have ranged from 0.1 to 3.9 percentage points below rates in the United Kingdom (International Monetary Fund, 1975). Differences in capital costs within countries are of even less importance than those between countries. This suggests that the cost of fuel is likely to be a more important determinant of international and interregional differences in the adoption of large-scale generating units than the cost of capital.

There is, however, one factor that does create significant differences in capital costs between utilities. In the United States interest on state or municipal bonds is not subject to federal taxation so investors are willing to lend money to utilities owned by these public bodies at lower interest rates than is the case for private companies. As a result, regions with many publicly owned utilities may have less incentive to adopt large generating units than other regions.

This exception, however, is related to the nature of ownership of electric utilities, and there are other reasons, which to some extent are offsetting, for expecting differences in the rate of adoption on the basis of ownership. For example, since large generating units reduce the capital needed for electric power production, privately owned utilities may be reluctant to introduce such innovations because their profits are often restricted by government regulation to a percentage return on their capital base. On the other hand, managers of private firms may be able to convert more of the savings realized from introducing large generating units into personal remuneration than is the case for managers of publicly owned facilities (Tilton, 1973). If so, the former should be willing to accept the risks associated with large units earlier than the latter. In addition, publicly owned firms may be subjected to greater pressure to buy domestic equipment even though larger, more efficient generating units are available from foreign equipment manufacturers. Thus there are *a priori* reasons for believing that public ownership stimulates the adoption of large generating units and *a priori* reasons for believing just the opposite. How the nature of ownership affects adoption in practice is uncertain.

Another factor that may influence the adoption of large generating units is the spatial distribution of demand for electric power. Although the average cost of generating electricity decreases as the size of generating units increases over the entire range of available equipment sizes, when the cost of transmission is taken into account, average cost may reach a minimum at a capacity level below that of the larger available units. This is particularly likely in remote or low-density markets, and so in such regions one would expect to find smaller generating units being introduced.

In areas where the base load of electricity is supplied by hydroelectric or nuclear generation, fossil-fueled generating capacity fills an intermediate or peaking function in meeting demand. Cycling operation of fossil-fueled units can be used to meet intermediate, often daytime, demand. However, this type of operation is not well suited to very large generating units which are designed for continuous base-load operation. Frequent starting and stopping with these large machines causes more frequent breakdowns and so necessitates the addition of more reserve capacity as well as causing higher maintenance costs for equipment in service (U.S. Federal Power Commission, 1971, p. I–5–7). In addition, the potential savings in fuel costs are reduced by this type of operation. This suggests that the extent to which a company or region relies on fossil fuels to generate its electric power may also affect the size of its fossil-fueled generating units.

Reliability

To meet demand in the event of the failure, or forced outage, of a generating unit, utilities maintain a certain amount of reserve capacity. Reserve capacity includes "spinning" reserves, which are units in operation and ready to be connected immediately to the system if needed, and standby reserves, which must be started up before they can carry any load. Factors considered by utility systems and power pools in determining their reserve capacity include the peak load, the size of the largest unit in service, and scheduled and unscheduled outage rates for units in the system.

Generally, firms have found reserve capacity equal to 15–25% of the annual peak load provides a level of reliability that on the average permits demand to exceed available capacity only once in a 10- or 11-year period. If the level of reserves decreases, reliability of service will deteriorate because less backup capacity is available to replace units forced out of service and existing capacity must be operated for longer periods, presenting more maintenance problems (Phillips, 1974, p. 1).

The failure of a large unit is less serious if the unit is part of a large system and so represents a small portion of total system capacity. For this reason, rarely will firms introduce a unit whose capacity exceeds 15% of the firm's total capacity. Thus increases in system size promote the introduction of larger generating units, though at some point the positive influence of system size diminishes simply because there is an upper limit on the available size of generating units at any particular time.

The largest generating units in service in 1974 were the 1300-MW units of the Tennessee Valley Authority and the American Electric Power Company in the United States. Firms with less than 7500 MW of capacity cannot seriously consider introducing units of this size, and in 1972 there were only nine electric power companies (excluding holding companies) in the United States and two in Canada whose total capacities exceeded 7500 MW.[2] The British Central Electricity Generating Board with over 60,000 MW of capacity is much larger than Canadian or United States utility companies. This suggests that small firm size should be an important factor inhibiting the adoption of large generating units in the United States and Canada, but not in Britain.

The variation in sizes of electric utilities in these three countries is the legacy of mergers, acquisitions, and government regulation described in the preceding chapter. In some areas of sparse population, small power companies are also the result of limited demand.

Several factors can to some extent offset the constraining effect of small firm size. These include interconnection agreements with neighboring utilities, joint ownership agreements for new generating facilities, and large growth rates of electricity demand.

The scope of interconnection agreements varies from the integrated operation of two or more electric power producers by a utility holding company to looser agreements between independent utilities to provide assistance during emergencies. Some holding companies such as the American Electric Power Company and tight pooling agreements such as the CAPCO group of utilities in the United States act effectively as one company in planning capacity additions. Such agreements reduce the disadvantages of small firm size by increasing the effective capacity of the electricity supply system. In this manner, system interconnections decrease the risk associated with the failure of large units.

The effect that interconnection agreements have on the size of generating units selected for capacity additions depends on the extent to which the agreement enables one company to rely on reserve capacity of another system. Some parties to interconnection agreements jointly plan their capacity additions. By "staggering" the installation of new units and selling excess power to another company until their demand increases sufficiently to justify a very large unit, they are able to install larger generating sets than would otherwise be possible.

For utilities to exploit interconnections with neighboring systems, adequate transmission facilities must exist to effect system links. In some areas, the remote nature of the system due to surrounding areas of sparse demand or natural boundaries has not stimulated construction of an adequate transmission network.

Small utilities can also install larger units than their size would justify by entering into joint ownership agreements with other companies. These agreements can cover complete power plants or individual generating units within a plant. Companies may also create a jointly owned subsidiary company for which capacity additions can be planned on a coordinated basis by the owners. This type of agreement may be more complex from a legal standpoint since the establishment of a new company may involve jurisdiction from more federal regulatory bodies than does joint ownership of a plant or unit.

Joint ownership has advantages over interconnection agreements in stimulating adoption of scale-related advances in generating equipment. The flexibility possible with joint ownership of production facilities affords an opportunity for small electric companies to add needed capacity to their system without being forced to install small units which do not provide all the economies possible in larger units or to enter agreements to purchase power from another generating company.

Interregional and international differences in the growth rate of electricity demand may also diminish the constraining effect of small firm size in areas where growth is rapid. A system with a rate of capacity growth higher than another system of similar size will be able to install larger units as the units will constitute a smaller percentage of future system capacity.

Three factors—interconnections, joint ownership, and growth rate of electricity demand—are expected to affect utility adoption behavior in areas where small firm size prevents the installation of large units due to the increased effect of unit failure on system reliability. Variations in these factors in regions with small system sizes may help explain differences in the rate at which large generating units are introduced.

Technical uncertainty has been found in other diffusion studies to be an important determinant of the rate at which new products or processes are adopted. If early adopters of an innovation find that unexpected problems obviate expected economic advantages, the industry may back off from the new technology and return to older, proven technology. Figure 1 presents some evidence to suggest that among the very largest units being introduced this problem did exist. The 1027-MW unit installed by the Consolidated Edison Company of New York in 1965 was 46% larger than the largest unit previously installed. The operating record of this unit was not good and it was five years before another increase in unit size occurred, this time only 12%. Overall, however, the influence of technical uncertainty does not appear to have been a major factor influencing the adoption of generating equipment during the period of this study since, as noted earlier, only a small percentage of units introduced were large enough to be on the technological frontier.

This chapter suggests that differences in fossil fuel costs and system size are major reasons for expecting the effect of large advanced steam-electric generating units on the costs and reliability of electric power to differ from one area to another. This expectation, in turn, leads to the hypothesis that the diffusion of new large-scale generating units should proceed more rapidly in those regions and countries where both fuel costs and system size are relatively great. Though probably of lesser importance, the nature of ownership, the extent of nonfossil-fueled generating capacity, market density, interconnections, joint ownership, and growth of electricity markets may also influence the pace of diffusion, at least in certain areas.

4
Diffusion in the United States

The first section of this chapter examines differences in fossil fuel costs and average system size for the nine geographic regions of the United States identified in Figure 3. On the basis of this information, those regions that would be expected to be rapid adopters of large fossil steam units and those that should be relatively slow are identified. The second section investigates the actual adoption behavior in these nine regions and assesses the extent to which expectations developed in the first section are borne out in practice. A simple econometric model is introduced that relates the average size of generating units introduced in a region to its

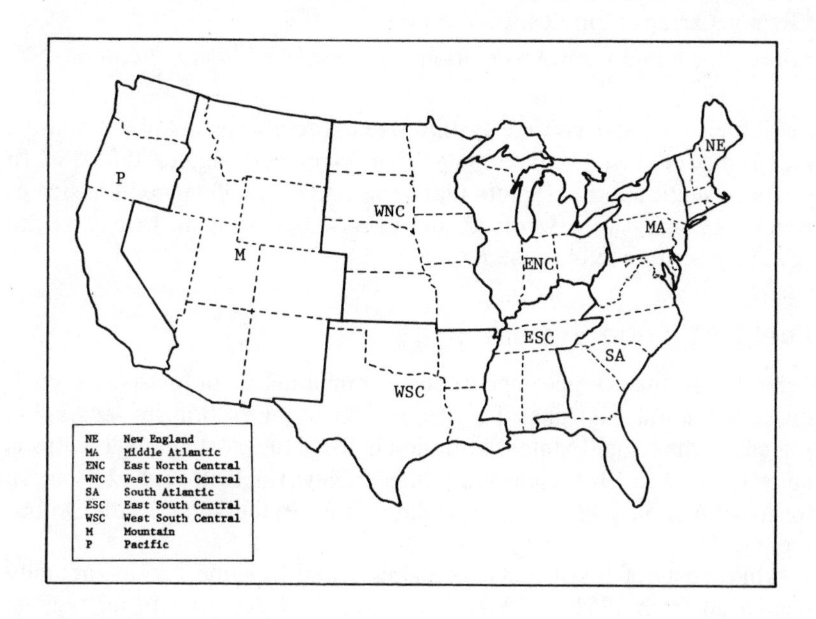

Figure 3. Geographic divisions of the United States.

Table 12
Average Cost of Fossil Fuel Used for Steam-Electric Generation by U.S. Electric Utilities by Region, 1952–72 (cents/million Btu[a])

	Year				
Region[b]	1952	1957	1962	1967	1972
New England	36.61	42.83	35.69	32.15	55.02
Middle Atlantic	29.64	33.10	30.51	29.57	51.31
Pacific	25.02	33.34	34.55	30.90	48.52
South Atlantic	27.06	31.21	27.37	27.70	44.42
East North Central	25.41	25.73	24.98	24.82	40.60
West North Central	24.75	25.01	25.35	24.93	33.13
East South Central	18.48	19.62	20.26	20.48	32.86
Mountain	18.86	22.39	26.86	23.01	28.75
West South Central	9.50	12.92	19.52	19.91	25.05
Total contiguous United States	24.89	27.06	26.48	25.71	39.90

[a] Current cents.

[b] Regions arranged in descending order of 1972 cost.

Source: National Coal Association, *Steam-Electric Plant Factors.*

fossil fuel costs and average system size in order to estimate the relative importance of these two factors on the speed of technological diffusion. In the third section, other factors that appear to have influenced the size of generating units are examined and an attempt is made to assess the relative importance of these factors.

Fuel Costs and System Size

Table 12 lists the average cost in cents per million Btu for fossil fuels (coal, oil, and natural gas) used for steam-electric generation in geographic regions of the United States. Variation in fossil fuel cost among regions is considerable. In 1972, electric utilities in New England paid 120% more for fossil fuel on a Btu basis than did utilities in the West South Central region.

While costs of fossil fuels rose slowly and in some regions actually decreased from 1952 to 1967, sharp increases occurred in all regions between 1967 and 1972 due largely to the switch to low-sulfur oil as a boiler fuel in many areas. The trend of rapidly rising fuel costs has continued since 1972. Those regions that depend on oil for utility boiler fuel are the

Table 13

Average Firm Size of Electric Utilities in the United States by Region, 1952–72 (MW)

Region	Year				
	1952	1957	1962	1967	1972
New England	86.46	113.22	162.28	200.14	343.04
Middle Atlantic	327.88	523.55	834.43	1,164.09	2,072.39
East North Central	183.73	338.56	426.58	478.71	721.52
West North Central	64.78	126.74	159.23	159.58	230.33
South Atlantic	188.64	314.09	522.82	625.65	1,100.22
East South Central	333.71	833.61	1,236.70	1,357.80	1,691.05
West South Central	167.32	317.76	495.82	589.89	919.28
Mountain	114.97	186.62	298.47	287.69	371.67
Pacific	315.50	504.60	752.82	990.45	1,182.83
Total contiguous United States	178.36	309.08	446.51	507.81	757.16

Source: U.S. Federal Power Commission, *Statistics of Privately Owned Electric Utilities in the United States*; U.S. Federal Power Commission, *Statistics of Publicly Owned Electric Utilities in the United States*; U.S. Department of Agriculture, Rural Electrification Administration, *Annual Statistical Report, Rural Electric Borrowers.*

first affected by higher prices for imported petroleum, but subsequent increases in the prices of alternative fuels also raised costs in coal-burning regions. The West South Central region, which includes Louisiana, Texas, Oklahoma, and Arkansas, relies on natural gas for many of its power plants. The cost of fuel in this region tends to lag behind that in other regions. One reason for this is that long-term contracts covering intrastate gas sales to electric utilities often do not have adjustment clauses, as do contracts covering other fuels that tend to prevail in other regions.

The relative costs of fossil fuels by region in the United States remained fairly consistent between 1952 and 1972. Four regions (New England, Middle Atlantic, Pacific, and South Atlantic) were always above the national average and three regions (West North Central, East South Central, and West South Central) were always below it. The Mountain states were only above the national average for one year of the five, and the East North Central States were below the national average for two of the five years.

The average size of electric utilities for various regions of the United States is seen in Table 13 for the 1952–72 period. For the purpose of analyzing the size of new generating units, however, it is the weighted average firm size, given in Table 14, that is appropriate. For example, if a

Table 14
Weighted Average Firm Size of Electric Utilities in the United States by Region, 1952–72[a] (MW)

	Year				
Region	1952	1957	1962	1967	1972
New England	308	462	634	964	1,489
Middle Atlantic	1,560	1,956	3,067	3,878	4,998
East North Central	900	1,545	2,326	3,158	5,124
West North Central	272	647	897	1,234	2,034
South Atlantic	779	1,227	1,823	2,683	4,713
East South Central	1,995	6,420	7,370	10,022	10,090
West South Central	342	667	1,388	2,001	3,298
Mountain	699	1,116	1,084	1,632	1,843
Pacific	1,809	2,335	3,284	5,075	6,551
Total contiguous United States	1,101	1,998	2,717	3,749	4,946

[a] Weighted average firm sizes were calculated by the formula

$$WFSZ = \sum_{i=1}^{n} \left(\frac{C_i}{\sum_{i=1}^{n} C_i} \right) C_i = \frac{\sum_{i=1}^{n} C_i^2}{\sum_{i=1}^{n} C_i}$$

where *WFSZ* is the weighted average firm size, C_i is the capacity in megawatts of the *i*th firm in the region, and the term in parentheses is the weighting factor (the firm's share of regional capacity).

Source: U.S. Federal Power Commission, *Statistics of Privately Owned Electric Utilities in the United States*; U.S. Federal Power Commission, *Statistics of Publicly Owned Electric Utilities in the United States*; U.S. Department of Agriculture, Rural Electrification Administration, *Annual Statistical Report, Rural Electric Borrowers*.

region contained one dominant firm that controlled 90% of the region's capacity and output and a number of small firms that together accounted for the remaining 10%, the average size of units introduced in that region should, all other things being equal, be relatively large, since the dominant

Table 15

United States Geographic Regions Grouped by Fossil Fuel Costs and Weighted Average Firm Size[a]

Weighted Average Firm Size	Average Fossil-Fuel Cost: High	Medium	Low
Large	Pacific Middle Atlantic		East South Central
Medium			
Small	New England South Atlantic	East North Central	West North Central West South Central Mountain

[a]Regions are assigned to the High or Large (Low or Small) categories if they were greater than (less than) the national average for four or more of the five years (1952, 1957, 1962, 1967, and 1972) studied. The Medium class includes those regions which were neither above nor below the national average for more than three years.

firm would install some 90% of all the new capacity in the region. Yet if the small firms in the region were numerous, the average firm size (but not the weighted average) would be small.

According to Table 14, the East South Central region, which contains the Tennessee Valley Authority, the nation's largest utility, is the region with the largest weighted average firm size in the United States. In 1972, its weighted average firm size of 10,090 MW was 54% greater than the next largest region (the Pacific states) and almost seven times as large as the New England region, which at 1489 MW was the smallest region.

The ranking of regions by weighted average firm size, like the ranking by fuel costs, has remained fairly consistent between 1952 and 1972. Two regions (East South Central and Pacific) have been above and five regions (South Atlantic, West South Central, West North Central, Mountain, and New England) have been below the average for the United States for the five years studied. The Middle Atlantic region has been above and the East North Central region below the national average for four out of the five years.

Chapter 3 indicates that large generating units will be introduced more rapidly in regions where fossil fuel costs are high and utility systems are large. Table 15 groups the United States geographic regions according to their average fossil fuel cost and weighted average firm size. Regions were assigned to the highest categories (most favorable to rapid adoption) if the fuel cost or firm size variable exceeded the national average for four or more of the five years examined in Tables 12 and 14. If the fuel cost or firm size variable was less than the national average for four or more of the five years, the region was assigned to the lowest categories (least favorable for rapid adoption). Other cases were assigned to an intermediate or medium category.

Under the assumption that fossil fuel costs and system size are the major variables affecting technological diffusion and that their effect upon diffusion is roughly equal, the faster adopters of new fossil-fueled steam-electric generating units should be located in the upper left-hand corner of Table 15. The Pacific states and Middle Atlantic states fall into this category. The regions where adoption is expected to proceed at a slower rate are found in the lower right-hand corner. Three regions—the West North Central region, the West South Central region, and the Mountain region—occupy this category.

The determination of diffusion rates in the other four geographic regions depends on the relative importance of fuel costs and firm size as a determinant of adoption behavior. The New England, South Atlantic, and perhaps East North Central states should have rapid adoption records if fuel cost is the leading determinant of adoption. If system size is the dominant factor, the East South Central region should be a fast adopter.

Adoption Behavior

Characteristics of new capacity additions to steam-electric generating capacity in the United States are given in the annual supplement to the U.S. Federal Power Commission's *Steam-Electric Plant Construction Cost and Annual Production Expenses*. For this study, information was collected on the rated nameplate capacity, turbine inlet steam pressure, and initial turbine inlet steam temperature for 1225 fossil-fueled steam-electric generating units larger than 10 MW installed by privately, publicly, and cooperatively owned utilities during the 20-year period from 1953 through 1972. The data for each unit were tabulated by the state in which the utility that owned the unit was listed in the FPC statistics, and were then aggregated into the nine geographic regions used by the Bureau of the Census (see Figure 3) in order to have a large enough sample for

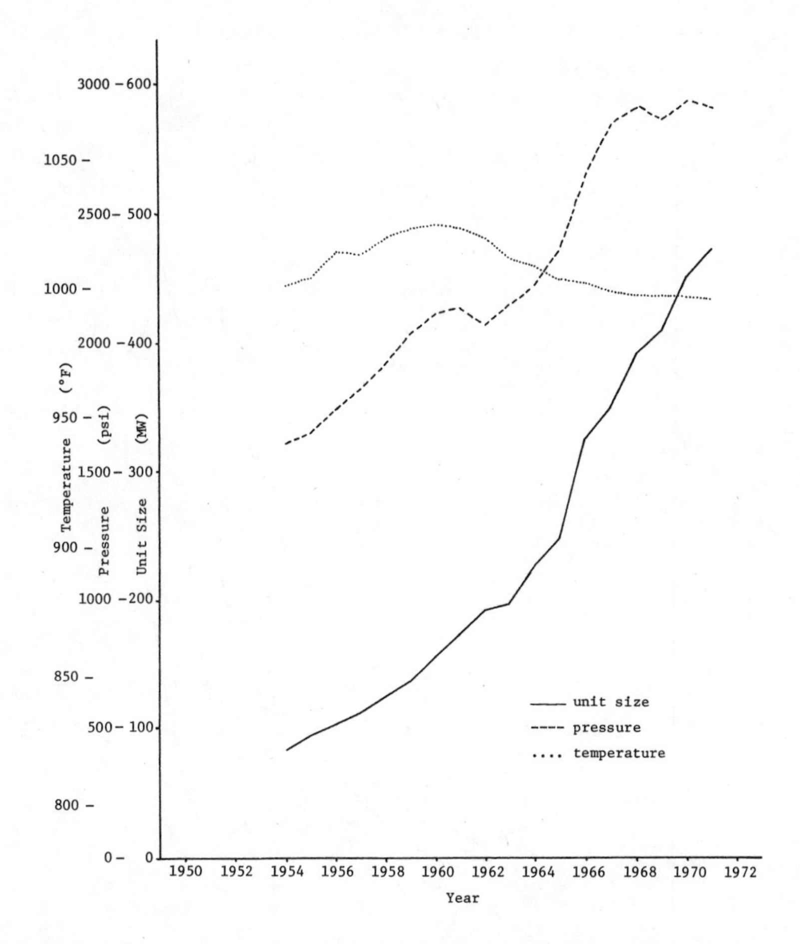

Figure 4. Three-year averages of nameplate capacity, inlet steam pressure, and inlet steam temperature for fossil-fueled steam-electric generating units introduced in the United States, 1954–71.

each year to calculate estimates of average size, pressure, and temperature.[1] Because some regions did not introduce any capacity during some of the years studied, averages of characteristics of generating units installed over a three-year period were used as a measure of technological advance of adopted units. The averages were designated by the central year of the three-year period.

Averages of unit size were calculated by dividing the sum of unit

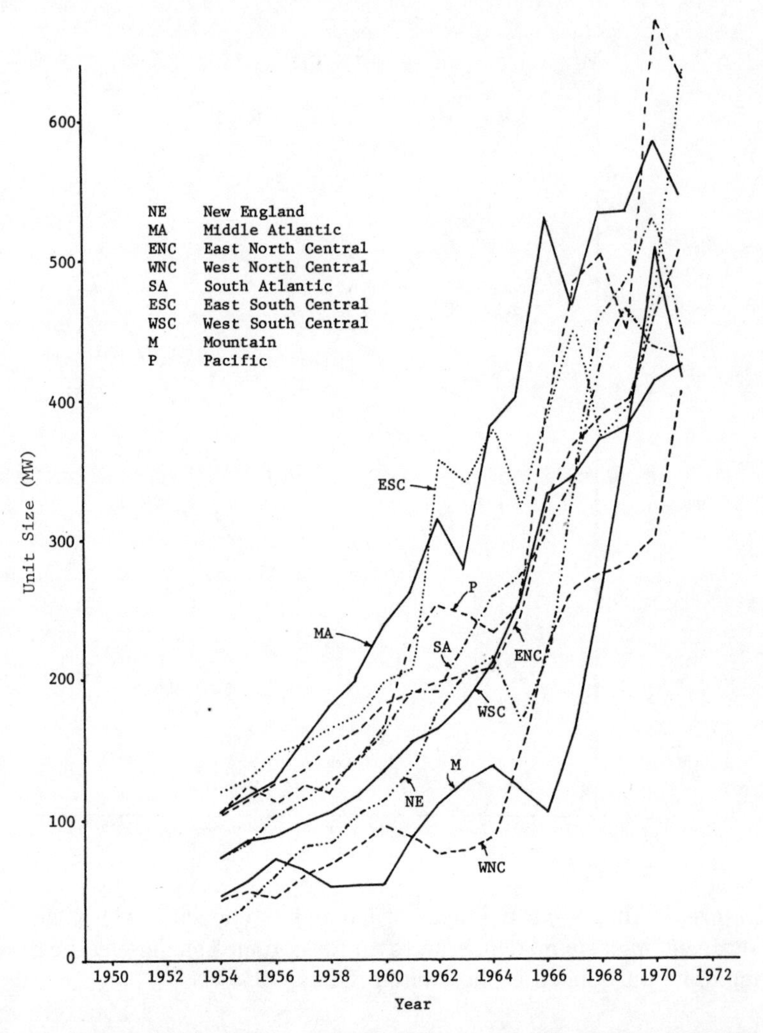

Figure 5. Three-year averages of nameplate capacity of fossil-fueled steam-electric generating units introduced in geographic regions of the United States, 1954–71.

capacities installed during the three-year period by the total number of units installed during that time. Averages of steam pressure were calculated by dividing the sum of the products of steam pressure times unit capacity for units introduced during the three-year period by the sum of the unit capacities installed. Averages of steam temperature were calculated in the same fashion as steam pressure.

Average size and pressure of units put in service in the United States from 1953 to 1972 have increased as shown in Figure 4.[2] Average steam temperature increased until it reached about 1030°F in 1960, but then declined to a level of about 1000°F in the early 1970s. The decline in temperature in recent years does not represent a reluctance on the part of utility management to utilize more advanced and efficient technology but rather a disenchantment with higher temperature units due to reliability and availability problems.

Three-year averages of size of units introduced in each geographic region from 1954 to 1971 are plotted in Figure 5. The regional increases in size of new generating units installed by utilities have followed a pattern similar to that of the national average. However, Figure 5 shows that certain regions, such as the Middle Atlantic states, have been early adopters, whereas other regions, such as the West North Central states, have installed relatively small units.

Interregional differences in unit size have been substantial. From 1965 through 1967, the average-sized unit installed in the Mountain region was 114 MW, about 21% of the 537-MW unit that represented the average size for the Middle Atlantic states during the same period. The time required for the region which installed the smallest average-sized units during a given period to attain an average unit size equivalent to that of the leading region of that initial period varies between five and eleven years.

To compare the actual adoption behavior with the behavior anticipated on the basis of regional system size and fuel cost, the average-sized generating units installed by regions are tabulated in Table 16 for 1957, 1962, 1967, and 1972. The 1972 figure is a two-year average for 1971 and 1972. The data for the other three years in the table are three-year averages centered on the nominal year.

This table also compares the average size of units introduced in the United States with Hughes' largest established scale class as listed in Table 3. The national three-year average for unit size is about 60% of the largest established scale class for the years shown in the table and fluctuates between 48 and 81% over the period 1954 to 1971. On a regional basis the ratio of average size to the Hughes' scale class varies over a larger range. From 1961 to 1963, for instance, the average unit installed in the East South Central region was 110% of the largest established scale class of 325 MW, whereas the average size of units installed in the West North Central region was only 24% of that scale class.

Regions are listed in Table 16 according to their average rank in installed

Table 16

Three-Year Averages of Nameplate Capacity of Fossil-Fueled Steam-Electric Generating Units Introduced in the United States by Region, 1957–72 (MW)

Region	Year[a] 1957	1962	1967	1972[b]	Average Rank
Middle Atlantic	160(1)[c]	319(2)	469(2)	557(2)	1.75
Pacific	123(4)	256(3)	486(1)	632(1)	2.25
East South Central	154(2)	359(1)	454(3)	553(3)	2.25
East North Central	135(3)	198(4)	370(4)	550(4)	3.75
South Atlantic	116(5)	.190(5)	339(6)	430(7)	5.75
New England	84(7)	178(6)	335(7)	445(6)	6.5
West South Central	101(6)	168(7)	347(5)	409(8)	6.5
West North Central	61(9)	78(9)	266(8)	474(5)	7.75
Mountain	66(8)	116(8)	165(9)	218(9)	8.5
Total contiguous United States	113	193	350	477	
Largest established scale class[d]	175	325	580	815	
Total contiguous United States as a percentage of largest established scale class	65	59	60	59	

[a] Central year of three-year average.
[b] Two-year average for 1971–72.
[c] Rank among geographic regions in parentheses.
[d] From Table 3.

Source: U.S. Federal Power Commission, *Statistics of Privately Owned Electric Utilities in the United States*; U.S. Federal Power Commission, *Statistics of Publicly Owned Electric Utilities in the United States*; U.S. Department of Agriculture, Rural Electrification Administration, *Annual Statistical Report, Rural Electric Borrowers.*

unit size for the four years tabulated. As was expected from the analysis of Table 15, the Middle Atlantic and Pacific regions lead the list as the earliest adopters and the West South Central, West North Central, and Mountain regions have introduced units that have been smaller than those of other regions. The East South Central region with low fuel costs but large firm size occupies the third position in adoption speed among the regions in Table 16. Regions where high to medium fossil fuel costs are coupled with small average system size (East North Central, South Atlantic, and New England) rank as moderate or slow adopters in Table 16. These findings suggest that system size has a greater influence on the diffusion of fossil-fueled steam-electric generating technology than the cost of fossil fuels.

A simple econometric model was also employed to assess systematically the relative effect of system size, fuel costs, and other variables. The

Table 17

Regression Results for Equation 1 Concerned with Regional Determinants of Average Unit Size in the United States, 1957–72[a]

Year		b_o	b_1	b_2	b_3	$\bar{R}^2$	n
1957	b_i	41.3	.321	.0534	-5.76×10^{-6}	.435	9
	(s_{b_i})	(37.9)	(1.149)	(.0245)	(3.45×10^{-6})		
	$\tilde{\beta}_i$		.016	.554	.430		
1962	b_i	−40.7	3.826	.0773	-4.56×10^{-6}	.778	9
	(s_{b_i})	(86.8)	(2.983)	(.0301)	(3.74×10^{-6})		
	$\tilde{\beta}_i$		.083	.617	.301		
1967	b_i	−36.1	8.402	.0763	-4.44×10^{-6}	.614	9
	(s_{b_i})	(149.2)	(5.766)	(.0350)	(3.19×10^{-6})		
	$\tilde{\beta}_i$		.095	.547	.359		
1972	b_i	89.9	4.802	.0609	-2.97×10^{-6}	.481	9
	(s_{b_i})	(140.8)	(3.039)	(.0458)	(4.01×10^{-6})		
	$\tilde{\beta}_i$		.164	.536	.300		

[a] Equation 1 is

$$Y_j = b_o + b_1F_j + b_2S_j + b_3S_j^2 + e_j, \quad j = 1, \ldots, 9$$

where Y_j = average size of generating units introduced in region j

F_j = fossil fuel cost in region j

S_j = weighted average firm size in region j

e_j = error term

relationship between average rated capacity of the turbine-generator units installed and regional determinants of adoption was analyzed with a multivariate regression model of the form

$$Y_j = b_0 + b_1X_{1,j} + \ldots + b_nX_{n,j} + e_j$$

where Y_j is the average capacity in megawatts of fossil-fueled steam-electric generating units installed in region j for a three-year period, $X_{1,j}, \ldots, X_{n,j}$ are the regional determinants of adoption behavior in region j, $b_0, \ldots, b_n$ are the regression coefficients, and e_j is the error term. The coefficients of the equation were estimated using data from the nine geographic regions of the United States. Equations for the years 1957, 1962, 1967, and 1972 were calculated and compared to see if the relative importance of the determinants of diffusion varied over time.

In the first equation estimated, the dependent variable Y_j is assumed to depend on the following:

1. The average cost F_j of fossil fuel (coal, oil, and natural gas) used for generation of electricity in region j. This variable, presented in Table 12, is measured in current cents per million Btu of the fuel's energy content.
2. The weighted average firm size S_j measured in megawatts of installed capacity of electric utilities weighted by the utility's percentage of total regional generating capacity, as seen in Table 14.
3. The square of weighted average firm size S_j^2. This variable is included because the stimulating effect that an increase in system capacity has on the size of new generating units should decrease after capacity reaches a certain size because there is an upper limit on the size of available generating units at any particular time.

According to the analysis in Chapter 3, the coefficients of F_j and S_j should be greater than zero and the coefficient of S_j^2 less than zero. The results of the regression model, presented in Table 17, indicate that the signs of the coefficients conform to these expectations for all four years. Although only the coefficients for firm size for 1957, 1962, and 1967 are significantly greater than zero at the 95% probability level (using a one-tailed t-test), this is not surprising given that only nine observations are available for the analysis.

Table 17 also presents values of $\tilde{\beta}_i$, which provide some indication of the relative importance or contribution of the independent variables in explaining changes in the dependent variable. The use of the beta coefficient, which is simply the coefficient for an independent variable

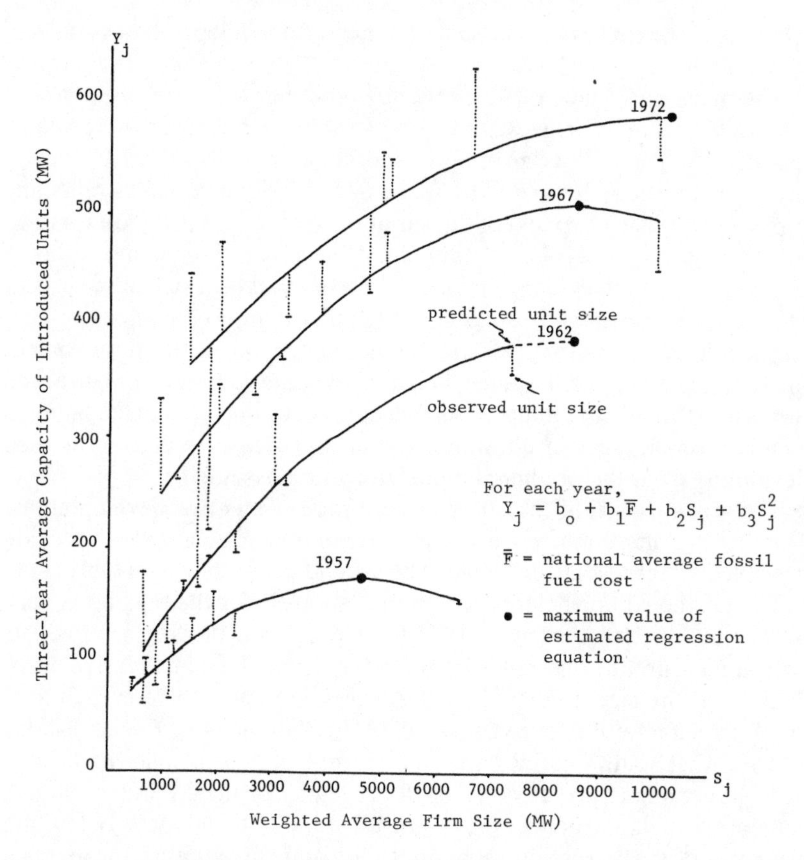

Figure 6. Relationship between three-year average of unit size and weighted average firm size in the United States, 1957–72.

multiplied by the ratio of its standard deviation to the standard deviation of the dependent variable, is described by Goldberger (1964, pp. 197–98). To show the fraction of the value of all beta coefficients represented by a beta coefficient for any one of the years modeled, the statistic

$$\tilde{\beta}_i = \frac{|\beta_i|}{\sum_{i=1}^{n} |\beta_i|}$$

where β_i is the ith beta coefficient and there are n independent variables, was calculated.

The increasing value of $\tilde{\beta}_1$ for the fuel cost variable over the 1957–72 period, for example, suggests that this variable has become a more important determinant of the size of new generating units over time. However, the small size of $\tilde{\beta}_1$ compared to $\tilde{\beta}_2$ and $\tilde{\beta}_3$ for the two variables reflecting firm size suggests that most of the variation in average unit size explained by the regression can be attributed to the effect of firm size. This is consistent with the evidence presented earlier that firm size has a greater influence on the size of new generating units than do fuel costs.

The average unit size predicted by the regression equation for each of the four years is plotted against weighted average firm size in Figure 6 on the assumption that regional fossil fuel costs corresponded to the national average for the year in question. Actual unit sizes are also shown, so deviations from the predicted unit size can be assessed.

The upward shift of the curves over time reflects two developments. The first and more important is the increase in available unit sizes made possible by technological change. The second is the change in fuel costs.

The corrected coefficients of determination $\bar{R}^2$ for the equations vary between 77.8% for 1962 and 43.5% for 1957. Table 18 lists the residuals (which measure the difference between the actual average unit size introduced and the size predicted by the regression equation) for each geographic region for the four years studied. In both the New England states and the West South Central states, the average sizes of installed units have been larger than predicted for all four years. In the Mountain states, generating units introduced have ranged from 24 to 112 MW below the estimates of the regression equation. In the other six regions, the average size of fossil-fueled steam-electric generating units introduced has been both above and below the predicted value.

These residuals arise because factors other than fuel costs and firm size influence diffusion, and their effect is not considered by the regression model. The next section looks at a number of factors identified in Chapter 3 as possible determinants of diffusion and tries to assess the influence they have had on the adoption of generating units in the United States.

Table 18

Residuals ($Y_j - \hat{Y}_j$) from Regression Equation for Average Unit Size by Region and Year

	Year							
	1957		1962		1967		1972	
Region	MW[a]	%[b]	MW	%	MW	%	MW	%
New England	5.6	6.6	34.6	19.5	31.5	9.4	7.2	1.6
Middle Atlantic	25.2	15.8	48.8	15.3	27.3	5.8	−9.1	−1.6
East North Central	16.8	12.4	−11.6	−5.8	1.1	0.3	31.0	5.6
West North Central	−20.1	−32.7	−44.3	−57.0	5.4	2.0	113.8	24.0
South Atlantic	7.3	6.3	0.1	0.0	−30.8	−9.1	−94.0	−21.9
East South Central	0.3	0.2	0.3	0.1	−0.1	−0.0	−5.7	−1.0
West South Central	22.2	22.1	36.0	21.4	81.8	23.4	30.2	7.4
Mountain	−35.4	−53.9	−24.1	−20.7	−105.0	−63.7	−111.8	−51.2
Pacific	−21.9	−17.7	−39.8	−15.5	−10.6	−2.2	38.4	6.1

[a] Residual $\Delta Y_j = Y_j - \hat{Y}_j$ in megawatts of nameplate capacity.

[b] Percent error = $\Delta Y_j / Y_j \times 100$.

Other Factors Affecting Diffusion

Interest Rates

Utilities located in regions with low interest rates have lower capital costs and do not benefit as much from the capital savings associated with large generating units as those utilities forced to pay high interest rates. Annual fixed charge rates which the U.S. Federal Power Commission estimates utilities will have to pay in 1990 are shown for National Power Survey regions in Table 19. Fixed charge rates include depreciation, insurance, and taxes, as well as the cost of capital, though the latter is probably the major component. In an illustration provided by the FPC, for example, the cost of money accounts for 58% of the fixed charge rate (U.S. Federal Power Commission, 1971, p. I–19–6). Still, the possibility cannot be completely ruled out that the regional differences in the fixed charge rates shown in Table 19 are due to factors other than differences in the cost of money.

With this caveat in mind, one does find that the northeastern United States with the highest fixed charge rate contains the New England and Middle Atlantic regions; according to Table 18 these regions introduced larger units than expected on the basis of their average firm size and fuel costs alone. In addition, the lowest fixed charge rate is found in the southeastern United States, which contains the South Atlantic region where smaller than expected units were installed. And the West, with the

Table 19

Estimated 1990 Annual Fixed Charge Rates for Fossil-Fueled Generating Equipment

Region[a]	Annual Fixed Charge Rate[b] (%)
Northeast	15.4
East Central	14.9
Southeast	12.7
West Central	14.3
South Central	14.2
West	13.9

[a] U.S. Federal Power Commission National Power Survey regions.

[b] Includes cost of money, depreciation and replacements, insurance, income taxes, and other taxes.

Source: U.S. Federal Power Commission, 1971, p. I–19–6.

second smallest fixed charge rate, includes the Mountain and Pacific regions where smaller than expected units were installed as well.

These findings suggest that regional differences in the cost of capital may influence the speed with which large generating units are adopted. Unfortunately, the evidence is at best tenuous due to the aggregate and imprecise nature of the data. To overcome this problem attempts were made to use bond ratings of utilities by region as a measure of the cost of capital, but they were unsuccessful. Thus before the importance of interest rates can be reliably appraised, better data on differences in the cost of capital and a finer regional breakdown than used in this study are needed.

Other Types of Generation

To estimate the effect of other types of generation, the regression model of Table 17 was expanded to include a variable C_j reflecting fossil steam's share of regional capacity. Data for this variable are presented in Table 20. The results of this expanded model, seen in Table 21, indicate that the

Table 20

Fossil-Fueled Steam-Electric Capacity as a Percentage of Capacity in the United States by Region, 1952–72

	Year				
Region	1952	1957	1962	1967	1972
New England	77.86	77.28	79.60	80.76	64.59
Middle Atlantic	88.68	91.86	90.70	80.93	71.92
East North Central	94.76	97.23	97.28	95.61	84.69
West North Central	84.09	80.88	83.56	76.38	76.47
South Atlantic	76.42	80.89	87.02	86.49	79.94
East South Central	44.24	68.84	76.37	79.76	81.07
West South Central	90.66	91.51	93.75	92.63	92.37
Mountain	30.18	40.17	48.16	47.91	53.37
Pacific	35.13	41.93	43.44	50.02	48.34
Total contiguous United States	73.90	78.38	80.88	79.39	75.28

Source: U.S. Federal Power Commission, *Statistics of Privately Owned Electric Utilities in the United States*; U.S. Federal Power Commission, *Statistics of Publicly Owned Electric Utilities in the United States*; U.S. Department of Agriculture, Rural Electrification Administration, *Annual Statistical Report, Rural Electric Borrowers*.

Table 21

Regression Results for Equation 2 Concerned with Regional Determinants of Average Unit Size in the United States, 1957–72[a]

Year		b_o	b_1	b_2	b_3	b_4	$\bar{R}^2$	n
1957	b_i	−46.6	.513	.0638	-6.91×10^{-6}	.954	.861	9
	(s_{b_i})	(28.7)	(.572)	(.0124)	(1.73×10^{-6})	(.236)		
	$\tilde{\beta}_i$		.019	.503	.392	.085		
1962	b_i	−197.0	6.043	.0740	-3.79×10^{-6}	1.241	.830	9
	(s_{b_i})	(123.7)	(2.952)	(.0264)	(3.30×10^{-6})	(.777)		
	$\tilde{\beta}_i$		.123	.555	.235	.087		
1967	b_i	−250.2	10.288	.0774	-4.43×10^{-6}	2.099	.744	9
	(s_{b_i})	(166.7)	(4.806)	(.0285)	(2.60×10^{-6})	(1.118)		
	$\tilde{\beta}_i$		.104	.496	.320	.081		
1972	b_i	−24.9	5.632	.0537	-2.43×10^{-6}	1.367	.394	9
	(s_{b_i})	(263.3)	(3.631)	(.0512)	(4.45×10^{-6})	(2.559)		
	$\tilde{\beta}_i$		.197	.485	.252	.066		

[a] Equation 2 is

$$Y_j = b_o + b_1F_j + b_2S_j + b_3S_j^2 + b_4C_j + e_j, \qquad j = 1, \ldots, 9$$

where Y_j = average size of generating units introduced in region j

F_j = fossil fuel cost in region j

S_j = weighted average firm size in region j

C_j = percentage of total capacity that is fossil-fueled steam-electric capacity in region j

e_j = error term

coefficient of C_j has the predicted positive sign for all four years. The explanatory power of the model as measured by the corrected coefficient of determination $\bar{R}^2$ shows improvement over the original model for three of the years. In 1972, however, the addition of the C_j variable causes a decrease in $\bar{R}^2$ from 48.1 to 39.4%. Moreover, the standard error of the C_j variable is far larger for 1972 relative to its coefficient than for earlier years, indicating that for 1972 at least the coefficient for this variable does not differ significantly from zero.

These findings suggest that the composition of generating capacity, at least until recently, has influenced the size of generating units introduced in the United States. The relative size of beta coefficients, though, indicates that this variable has had much less of an effect on the average size of generating units than firm size. And only in 1957 was this variable more influential than fuel cost.

Spatial Distribution of Demand

Transmission costs may inhibit the addition of large generating units in regions where the markets for electricity are sparse. The areal density of electricity demand in regions of the United States can be approximated by population density.

Population densities, measured in people per square mile, for the nine geographic regions are available from the 1960 and 1970 census. Since the differences between regions in population densities remained nearly the same between the two census years, an average of the 1960 and 1970 population densities was used as a measure of interregional difference in the spatial distribution of electricity demand for all of the four years studied. Areas of low population density, where smaller than average units are likely to be adopted, are the Mountain, West North Central, and West South Central regions.

The results of the addition of this variable, P_j, to the firm size-fuel cost model are presented in Table 22. The sign of the coefficient of P_j was positive as expected for all years except 1972 and significantly different from zero for 1957 and 1962. Again, however, the $\tilde{\beta}_i$ statistics indicate that the effect of firm size has been far greater than either population density or fuel cost. The explanatory power of the regression equation as measured by the corrected coefficient of determination $\bar{R}^2$ increased when population density was added to the 1957 and 1962 equations but decreased for the 1967 and 1972 equations. The decreasing trend of $\tilde{\beta}_4$ over the period 1957–72 indicates that the effect of demand density on diffusion of scale-

Table 22

Regression Results for Equation 3 Concerned with Regional Determinants of Average Unit Size in the United States, 1957–72[a]

Year		b_o	b_1	b_2	b_3	b_4	$\bar{R}^2$	n
1957	b_i	68.4	−.950	.0374	-3.65×10^{-6}	.209	.773	9
	(s_{b_i})	(25.8)	(.850)	(.0165)	(2.30×10^{-6})	(.072)		
	$\tilde{\beta}_i$		.057	.465	.327	.151		
1962	b_i	16.5	1.670	.0530	-1.90×10^{-6}	.305	.897	9
	(s_{b_i})	(63.1)	(2.197)	(.0226)	(2.75×10^{-6})	(.118)		
	$\tilde{\beta}_i$		.051	.596	.177	.177		
1967	b_i	32.9	5.387	.0668	-3.71×10^{-6}	.246	.600	9
	(s_{b_i})	(170.1)	(6.752)	(.0371)	(3.35×10^{-6})	(.272)		
	$\tilde{\beta}_i$		.067	.528	.331	.074		
1972	b_i	85.2	4.943	.0615	-3.03×10^{-6}	−.021	.351	9
	(s_{b_i})	(190.1)	(4.660)	(.0533)	(4.64×10^{-6})	(.480)		
	$\tilde{\beta}_i$		.165	.530	.298	.007		

[a] Equation 3 is

$$Y_j = b_o + b_1F_j + b_2S_j + b_3S_j^2 + b_4P_j + e_j, \quad j = 1, \ldots, 9$$

where Y_j = average size of generating units introduced in region j

F_j = fossil fuel cost in region j

S_j = weighted average firm size in region j

P_j = regional population density (1960–70 average) in region j

e_j = error term

Table 23
Privately Owned Capacity as a Percentage of Total Capacity in the United States by Region, 1952–72

	Year				
Region	1952	1957	1962	1967	1972
New England	97.34	96.86	96.91	96.89	97.70
Middle Atlantic	98.94	99.11	96.15	91.32	93.92
East North Central	94.54	95.99	96.06	94.96	94.86
West North Central	82.30	80.85	74.45	66.71	70.38
South Atlantic	94.55	91.65	91.02	90.17	92.13
East South Central	39.60	27.33	33.25	32.26	40.14
West South Central	89.06	88.48	89.28	84.54	83.70
Mountain	53.77	50.60	59.20	58.09	62.51
Pacific	53.18	51.73	53.76	55.81	54.77
Total contiguous United States	82.38	78.64	79.09	76.69	79.62

Source: U.S. Federal Power Commission, *Statistics of Privately Owned Electric Utilities in the United States*; U.S. Federal Power Commission, *Statistics of Publicly Owned Electric Utilities in the United States*; U.S. Department of Agriculture, Rural Electrification Administration, *Annual Statistical Report, Rural Electric Borrowers.*

related technological advances in fossil-fueled generating units may not be as important now as it was in earlier years. This is not surprising in light of recent technological developments reducing the cost of transmitting bulk electric power over long distances.

Nature of Ownership

Regional differences in the ownership status of electric utilities in the United States are shown in Table 23 and Table 24. Some of the reasons for believing that the type of ownership of a utility is a factor in the determination of average unit size were considered earlier. Lower interst rates, for example, are available to publicly and cooperatively owned firms. There may also be differences in the abilities of managers and the incentives they have to introduce new technology associated with the nature of ownership.

In order to estimate the effect that ownership has on the size of fossil-fueled steam-electric generating units installed in a region, an ownership

Table 24

Privately and Federally Owned Capacity as a Percentage of Total Capacity in the United States by Region, 1952–72

Region	Year				
	1952	1957	1962	1967	1972
New England	97.34	96.86	96.91	96.89	97.70
Middle Atlantic	98.94	99.11	96.15	91.32	93.92
East North Central	94.62	96.05	96.06	95.00	94.88
West North Central	82.30	88.86	83.02	77.59	77.86
South Atlantic	95.29	95.82	94.29	93.01	93.83
East South Central	98.56	98.86	95.11	93.75	92.35
West South Central	91.91	93.56	92.33	88.23	86.68
Mountain	93.39	90.81	90.05	89.12	88.30
Pacific	83.01	81.81	79.37	75.05	75.64
Total contiguous United States	93.37	93.93	91.87	88.89	89.37

Source: U.S. Federal Power Commission, *Statistics of Privately Owned Electric Utilities in the United States*; U.S. Federal Power Commission, *Statistics of Publicly Owned Electric Utilities in the United States*; U.S. Department of Agriculture, Rural Electrification Administration, *Annual Statistical Report, Rural Electric Borrowers.*

variable was added to the firm size-fuel cost model of Table 17. Table 25 shows how O_j, the percentage of regional capacity owned by investor-owned utilities, influences the dependent variable, unit size. The sign of the coefficient of O_j was positive for 1957, 1962, and 1967 and then negative for 1972. Moreover, the effect of this variable has been small compared to firm size, and declining in importance at least since 1962 according to the $\tilde{\beta}_i$ statistics. Data from Table 24 indicating federally and privately owned capacity in a region were used as an alternative measure of O_j in this model. The results, however, were inconclusive.

Ownership of generating capacity by investor-owned firms seems to have been a minor positive factor in influencing adoption in the earlier years of study. However, the evidence is tenuous, and data aggregated on a regional basis are not the most appropriate for assessing the effect of the nature of ownership. Studies based on individual firm data presumably would provide a more definitive test of the relationship between diffusion and the nature of ownership. One such study (Tilton, 1973) does provide some support for the hypothesis that privately owned, regulated firms adopt cost-saving innovations more rapidly than publicly owned firms.

Table 25

Regression Results for Equation 4 Concerned with Regional Determinants of Average Unit Size in the United States, 1957–72[a]

Year		b_o	b_1	b_2	b_3	b_4	$\bar{R}^2$	n
1957	b_i	−50.5	−.315	.0643	-5.68×10^{-6}	1.173	.920	9
	(s_{b_i})	(21.8)	(.448)	(.0094)	(1.30×10^{-6})	(.210)		
	$\tilde{\beta}_i$		.012	.521	.332	.135		
1962	b_i	−185.0	4.332	.0685	-1.88×10^{-6}	1.642	.887	9
	(s_{b_i})	(85.9)	(2.137)	(.0218)	(2.88×10^{-6})	(.679)		
	$\tilde{\beta}_i$		.103	.600	.137	.160		
1967	b_i	−148.9	6.700	.0715	-3.09×10^{-6}	1.986	.677	9
	(s_{b_i})	(158.3)	(5.411)	(.0322)	(3.07×10^{-6})	(1.413)		
	$\tilde{\beta}_i$		.079	.539	.262	.119		
1972	b_i	106.5	5.036	.0635	-3.35×10^{-6}	−.362	.354	9
	(s_{b_i})	(201.8)	(3.828)	(.0548)	(5.29×10^{-6})	(2.757)		
	$\tilde{\beta}_i$		.157	.512	.309	.022		

[a] Equation 4 is

$$Y_j = b_o + b_1F_j + b_2S_j + b_3S_j^2 + b_4O_j + e_j, \quad j = 1, \ldots, 9$$

where Y_j = average size of generating units introduced in region j

F_j = fossil fuel cost in region j

S_j = weighted average firm size in region j

O_j = investor-owned utilities' percentage of capacity in region j

e_j = error term

Access to Equipment Manufacturers

According to Epstein (1972, p. 22), a number of utilities accounting for 16% of the generating capacity of the United States will not consider the purchase of imported equipment. Epstein does not explain why these utilities are committed to ordering American equipment, but she indicates that recently more utilities appear willing to consider imports. Surrey and Chesshire (1972, p. 81) suggest that utilities may prefer domestic suppliers because they are dependent on their suppliers to service existing facilities. In addition, the procurement of equipment by federally owned utilities falls under the jurisdiction of the Buy American Act, which requires purchases from U.S. manufacturers unless the lowest foreign bid, including import duty and costs of entry, is at least 6% below the lowest domestic bid. Moreover, if the domestic product is to be produced in an area of substantial unemployment, this differential becomes 12%. Finally, even if a foreign bid is more than 12% below the lowest domestic bid, utilities may accept a domestic bid for reasons of national interest, to assist small business firms, or to protect national security (Epstein, 1972, p. 24).

One foreign manufacturer that has sold several steam-electric generating units to utilities in the United States is Switzerland's Brown Boveri. In addition to the 1300-MW units built for the American Electric Power Company and the Tennessee Valley Authority in the early 1970s, Brown Boveri has sold smaller units ranging from 315 to 615 MW to other utilities including the Basin Electric and Power Cooperative, Buckeye Power Inc., Virginia Electric and Power Company, and the Los Angeles (California) Department of Water and Power (General Electric EUBP data bank, 1974).

Although government policies that restrict access to foreign equipment conceivably could inhibit the adoption of the largest unit since Brown Boveri has at times pioneered scale advances, their effect in practice appears to have been negligible for two reasons. First, General Electric and Westinghouse produce an extensive range of unit sizes of sufficient magnitude to meet the requirements of nearly all utilities in the United States. Second, in a number of instances where large foreign units were desired, imports occurred in spite of the Buy American Act and other restrictions as the recent acquisition of 1300-MW units just mentioned illustrates.

Interconnections

Although system size poses a constraint to all but the largest utilities desiring to install new units, this constraint can be alleviated by coordination between systems in the form of interconnections, pools, and regional reliability groupings. Interconnection agreements have been widely used in the United States where utility systems with established transmission networks have been close to similar systems. Power pools and interconnections differ in their scope but all such agreements serve to increase system reliability or to diminish the reserve capacity requirement associated with smaller system size. The formation of National Electricity Reliability Councils in the United States assisted electricity supply systems in coordinating their operations on a national basis.

Interconnection agreements presumably should stimulate the addition of larger units since utilities can draw on the reserve capacity of other systems in emergencies. But formulas for calculating the capacity obligations of participants may to some extent offset this beneficial effect. In the case of the Pennsylvania-New Jersey-Maryland Interconnection (PJM), for example, the capacity that each party is required to provide depends, among other things, on the number of units it possesses in excess of 900 MW and the amount by which its average FOR differs from the PJM rate. The latter, in turn, tends to be higher for larger units as pointed out in Chapter 2. The PJM agreement therefore may not stimulate the adoption of very large units to the extent anticipated (PJM Interconnection Agreement, 1 April 1974, Schedule 2.21).

Still, interconnections such as that of the PJM group, which coordinate the operations of member systems to minimize overall costs, encourage the adoption of larger units more than looser agreements that simply insure reliability of supply. Utilities which are a part of interconnection agreements that integrate the operations of member systems account for large amounts of regional capacity in the Middle Atlantic, South Atlantic, and West South Central states. Interconnections of this nature also include an appreciable amount of the capacity in the states of the East South Central, East North Central, and New England regions. Such interconnection agreements are less prevalent in the West North Central, Mountain, and Pacific regions.

The PJM Interconnection and Allegheny Power System include companies in both the Middle Atlantic and South Atlantic regions. Some other utilities in the South Atlantic region belong to the Southern Company System and the American Electric Power Company system. Texas Utilities Company, Central and Southwest Corporation, and Middle

Table 26

Multiple-Owned Fossil-Fueled Steam-Electric Plants in the United States, 1967–72

Plant	State	Year[a]	Plant Size[b] (MW)	Units[c]	Average Unit Size (MW)	Number of Companies
Keystone	PA	1967	1,872.0	2	936	7
Cardinal	OH	1967	1,230.5	2	615	2[d]
Fort Martin	WV	1967	1,152.0	2	576	4
Canal	MA	1968	542.5	1	542	4[e,f]
Four Corners (4 & 5)[g]	NM	1969	1,636.2	2	818	6
Homer City	PA	1969	1,319.4	2	660	2
Hatsfield Ferry	PA	1969	1,728.0	3	576	3[f]
W.C. Beckjord (6)	OH	1969	460.8	1	461	3
Edgewater (4)	WI	1969	351.0	1	351	2
Conemaugh	PA	1970	1,872.0	2	936	9
J.M. Stuart	OH	1970	1,830.6	3	610	3
Warrick (4)	IN	1970	380.0	1	380	2[h]
Mojave	NV	1971	1,636.1	2	818	4
W.H. Sammis (7)	OH	1971	800.0	1	800	3
Big Brown	TX	1971	1,186.8	2	593	3[f]
Harrison	WV	1972	684.0	1	684	3
Eastlake (5)	OH	1972	680.0	1	680	2
Bowline Point	NY	1972	621.0	1	621	2

[a] Initial year of operation of jointly owned plant or units.

[b] Nameplate capacity.

[c] Units included in jointly owned plants.

[d] One investor-owned utility and a corporation comprised of 28 nonprofit electric companies operating in Ohio on a cooperative basis.

[e] Canal Electric Co. is a wholesale generating company owned by four New England utilities or utility holding companies.

[f] All owners are subsidiaries of one holding company.

[g] Numbers in parentheses identify the jointly owned unit(s) in a plant that is not entirely jointly owned.

[h] Jointly owned by an electric utility and a subsidiary of Aluminum Company of America.

Source: U.S. Federal Power Commission, *Steam-Electric Plant Construction Cost and Annual Production Expenses*, 1972; General Electric Company, 1974.

South System are large holding companies serving the West South Central states.

The American Electric Power Company system, the Middle South System, and the Southern Company System own utilities in the East South Central states. Utilities of the American Electric Power Company also operate in the East North Central region. Several small investor-owned utilities in the New England region have become members of holding companies, such as Northeast Utilities and New England Gas and Electric Association.

Of the eight positive residuals in Table 18 which exceed 10% of the value of the dependent variable, five are for the Middle Atlantic and West South Central regions which have a large amount of capacity covered by system interconnection agreements; two are for the New England and East North Central regions which have a moderate amount of capacity involved in interconnection agreements; and one is for the West North Central region where formal system interconnections are less pervasive. Eight of the nine negative residuals in Table 18 which are larger in absolute value than 10% of Y_j occur for regions of low interconnection activity—the Mountain, Pacific, and West North Central regions. The only large negative residual occurring where interconnections are extensive is for the South Atlantic region for the year 1972.

Although this brief examination of interconnections looks only at larger holding companies and formal interconnection agreements, it appears that this factor explains some of the residuals from the firm size-fuel cost equation. However, the PJM formula for capacity obligations demonstrates that the constraining effects of system size are not completely removed by coordination agreements.

Joint Ownership Agreements

Joint ownership agreements, which have increased appreciably in number since 1967, provide small utilities with the opportunity to take advantage of economies of scale and still purchase only the additional capacity they need. One product of such an agreement is the Conemaugh mine-mouth plant near New Florence, Pennsylvania, which was constructed in 1970. Nine firms own this 1872-MW facility (two 936-MW coal-fired units) with equity shares ranging from Public Service Electric and Gas Company's 22.5% (421.2 MW) to United Gas Improvement Company's 1.11% (20.8 MW). The Conemaugh plant is only one of a number of plants built under joint ownership agreements. Others are listed in Table 26, though this list is not exhaustive. Joint owners may be parts of the same holding company (the three owners of the Hatsfield Ferry plant are all owned by the Allegheny Power System) or members of the same interconnection agreement (the nine owners of the Conemaugh plant belong to the Pennsylvania-New Jersey-Maryland Interconnection).

Utilities and utility holding companies may form a jointly owned wholesale generating company as the Canal Electric Company did in New England. The Cardinal plant in Ohio is jointly owned by Buckeye Power Inc., an association of Ohio's 28 rural electric cooperatives, and Ohio Power Company, an investor-owned electric power company that is a part of the American Electric Power Company system. Coordination of plant operation is achieved by a single operating company, Cardinal Operating Company, which is equally owned by Buckeye Power Inc. and Ohio Power Company (Sporn, 1968, pp. 445–46). Seven investor-owned utilities in New York State proposed forming a jointly owned corporation to build and operate generating units. The proposed corporation, Empire State Power Resources Inc., was to give the owners a stronger base upon which to finance future generating units and to alleviate some siting problems caused by the uneven distribution of available power plant sites among the service areas of owner utilities (*Wall Street Journal*, 4 April 1974, p. 16).

Two-thirds of the plants listed in Table 26 are situated in the Middle Atlantic and East North Central regions where sizes of new generating units have often been larger than predicted by the firm size-fuel cost equation. Since joint owners are often mutual partners in interconnection agreements or subsidiaries of the same holding company, it is difficult to separate the influence of joint ownership from these other considerations. Judging from the size of units installed under joint ownership agreements that are listed in Table 26, such agreements appear to enable companies to install larger units than would otherwise be possible. Given that most joint ownership agreements are of recent origin and that their numbers so far are not great, their effect has probably been modest in the past. However, it may increase substantially in the future.

Growth of Demand

Another factor that presumably fosters the introduction of larger units is rapid growth in demand. An area with a smaller amount of installed capacity than another area but a higher growth rate may install larger units because of a larger projected system size.

Table 27
Average Annual Regional Growth Rate in Generating Capacity in the United States over the Preceding Five-Year Period by Region, 1957–72 (%)

Region	Year			
	1957	1962	1967	1972
New England	5.14	5.31	4.28	10.41
Middle Atlantic	5.60	8.56	5.64	7.81
East North Central	9.54	7.82	4.38	8.14
West North Central	12.36	9.38	8.23	7.79
South Atlantic	9.01	10.27	8.75	10.83
East South Central	17.30	8.21	6.54	5.51
West South Central	13.69	13.02	10.09	10.09
Mountain	9.50	7.71	7.90	4.77
Pacific	9.21	9.58	7.34	6.14
Total contiguous United States	9.67	9.00	6.85	8.11

Source: U.S. Federal Power Commission, *Statistics of Privately Owned Electric Utilities in the United States*; U.S. Federal Power Commission, *Statistics of Publicly Owned Electric Utilities in the United States*; U.S. Department of Agriculture, Rural Electrification Administration, *Annual Statistical Report, Rural Electric Borrowers*.

The growth rates in capacity reported in Table 27 were used to estimate regional growth rates in demand. This variable was then used along with firm size and fuel cost to explain the average unit size of new fossil steam units. The results, presented in Table 28, indicate that the performance of the new variable is poor. For two years its coefficient has the wrong sign, and for all years it is insignificant. The importance of the new variable compared to firm size was small according to the beta coefficients, and its addition decreased the $\bar{R}^2$ for all four years. Apparently growth in demand has little or no effect on the size of new generating units.

In summary, by far the most important factor determing the size of new generating units in the United States appears to be firm size. Fuel costs, although anticipated by Chapter 3 to be of comparable importance, have a substantially smaller effect. The density of population, nature of ownership, and percentage of generating capacity run on fossil fuels may also influence the size of units. Though less important than firm size, these variables may have an effect equivalent to fuel costs. Coordinated plan-

Table 28

Regression Results for Equation 5 Concerned with Regional Determinants of Average Unit Size in the United States, 1957–72[a]

Year		b_0	b_1	b_2	b_3	b_4	$\bar{R}^2$	n
1957	b_i	157.0	−1.447	.0441	-3.63×10^{-6}	−6.283	.393	9
	(s_{b_i})	(148.5)	(2.493)	(.0279)	(4.44×10^{-6})	(7.780)		
	$\tilde{\beta}_i$		0.77	.488	.289	.145		
1962	b_i	−41.0	3.831	.0773	-4.56×10^{-6}	.017	.722	9
	(s_{b_i})	(208.8)	(4.913)	(.0402)	(5.25×10^{-6})	(12.331)		
	$\tilde{\beta}_i$		.083	.617	.300	.000		
1967	b_i	−21.9	8.108	.0767	-4.50×10^{-6}	−1.001	.518	9
	(s_{b_i})	(280.8)	(7.944)	(.0397)	(3.68×10^{-6})	(15.860)		
	$\tilde{\beta}_i$		.091	.544	.360	.005		
1972	b_i	68.5	4.626	.0592	-2.72×10^{-6}	3.676	.357	9
	(s_{b_i})	(190.1)	(3.495)	(.0516)	(4.63×10^{-6})	(18.517)		
	$\tilde{\beta}_i$		.161	.532	.280	.026		

[a] Equation 5 is

$$Y_j = b_0 + b_1 F_j + b_2 S_j + b_3 S_j^2 + b_4 G_j + e_j, \quad j = 1, \ldots, 9$$

where Y_j = average size of generating units introduced in region j

F_j = fossil fuel cost in region j

S_j = weighted average firm size in region j

G_j = average annual capacity growth rate for previous five years in region j

e_j = error term

ning of capacity additions through system interconnection and joint ownership agreements offsets the constraining effect of small firm size, but only to a certain extent. Little or no evidence was found that interest rates, growth of demand, and restrictions on foreign equipment purchases significantly affect the adoption of large generating units.

5 Diffusion in Canada

This chapter investigates the adoption of electric power generating units in Canada. The first section examines average fuel costs and system size in eastern and western Canada. On the basis of this information, it then indicates which of these two regions one would expect to introduce the more advanced units as well as how adoption in these two regions should compare with that of various regions in the United States. The second section looks at the actual adoption behavior in Canada and determines the extent to which it conforms to the expectations or hypotheses advanced in the first section. The third section identifies and assesses the relative importance of factors other than fuel costs and system size that influence the adoption of new generating units in Canada.

Fuel Costs and System Size

Costs of fossil fuels used for the generation of electricity in Canada are given in Table 29. Eastern Canada, which includes Newfoundland, Prince Edward Island, Nova Scotia, New Brunswick, Quebec, and Ontario, accounts for 66% of the national total of fossil fuels used for generation of electricity, and western Canada, which includes Manitoba, Saskatchewan, Alberta, and British Columbia, for 34%. The northern region covering the Northwest Territories and Yukon accounts for less than 1%. The low cost of natural gas used as utility boiler fuel in Alberta and Saskatchewan gives those provinces the lowest cost fuel in Canada at 12.56 and 18.41 cents/million Btu, respectively. This is similar to the situation in the United States where the West South Central region containing the gas-producing states of Texas and Louisiana has the lowest cost boiler fuel (see Table 12).

Oil used for Canada's electricity generation comes in about equal amounts from domestic sources (utilities in western Canada use oil from

Table 29

Average Cost of Fossil Fuel Used by Electric Utilities in Canada by Province and Region, 1972 (cents[a]/million Btu)

Region/Province	Cost
East	44.90
Newfoundland	66.92
Prince Edward Island	49.43
Nova Scotia	36.56
New Brunswick	39.96
Quebec	62.14
Ontario	46.45
West	16.57
Manitoba	41.53
Saskatchewan	18.41
Alberta	12.56
British Columbia	40.36
North[b]	168.24
Northwest Territories	165.47
Yukon	177.44
Total[c]	35.49

[a] 1972 Canadian cents.

[b] This region not included in study as fossil fuels are used primarily for generation by internal combustion engines.

[c] Includes all provinces and territories.

Source: Statistics Canada, 1974b, p. 36–39.

Alberta and Saskatchewan) and from the Middle East (used mainly in eastern Canada). Canada could be self-sufficient in oil supply, but half of the western Canadian oil production is exported to the United States. This situation may change if a projected trans-Canadian pipeline is constructed to bring petroleum from the western provinces to eastern Canada.

The federal government, at the recommendation of the National Energy Board,[1] established the National Oil Policy line in 1960, so that all oil used west of the Ottawa River (Ontario and west) must come from domestic sources and utilities east of the line may use imported oil. Under this policy, Alberta oil was exported to the United States in 1974 at a price of about $6.50/barrel plus an export tax which brought the price to the international level. Oil used in eastern Canada was subsidized at a price of

$6.50/barrel plus transportation costs from western Canada, which raised the price to about $7.50–8.00/barrel (A. R. Scott, Canadian Department of Energy, Mines and Resources, oral commun., 1974). Debanné (1974) argues that this policy merely formalized the existing policy of multinational oil companies which controlled operations in Canada and the Middle East.

About half of the coal burned in Canada to generate electricity is imported from the United States by the Hydro-Electric Power Commission of Ontario (Ontario Hydro). This imported coal accounted for 46.9% of the electricity generated by fossil fuels in Canada in 1972 (Statistics Canada, 1974b, pp. 38–39). Coal from fields in western Canada is used primarily by utilities in Alberta and Saskatchewan. Coal prices in Canada have risen to United States levels and a regional price difference exists due to transportation costs from western Canada. Coal which costs 10–15 cents/million Btu in Alberta may cost 50–55 cents/million Btu in Ontario (A.R. Scott, oral commun., 1974).

The 1972 regional average cost of fossil fuel used for electricity generation, as seen in Table 29, was 44.90 Canadian cents/million Btu in eastern Canada, compared with only 16.57 Canadian cents/million Btu in western Canada. Thus the stimulating effect of fuel costs on the adoption of large generating units should be greater for eastern Canada.

In U.S. currency, the average fossil fuel costs for eastern, western, and all of Canada were 44.70, 16.50, and 35.33 cents in 1972. Comparing these values with fossil fuel costs in the United States (Table 12), one finds fuel costs in Canada average 11% less than in the United States. Fuel costs in western Canada are lower than for any U.S. region, and the costs in eastern Canada are similar to those of the South Atlantic region, which ranked fourth in average fuel costs among United States regions in 1972.

In seven Canadian provinces, one provincially owned electric utility controls more than 90% of utility generating capacity for the province. Table 30 presents the capacities of these firms and large electric utilities in other provinces at the end of 1973. Systems range in size from those of Ontario Hydro and Quebec Hydro, which both exceed 10,000 MW and together account for 56.7% of utility generating capacity in Canada, to the 111-MW Maritime Electric Company, Limited, which supplies electricity to Prince Edward Island. Maritime Electric's capacity is less than 1% that of Ontario Hydro or Quebec Hydro.

On the basis of the systems listed in Table 30, the weighted average firm size of electric utilities in eastern and western Canada is 11,916 and 2807 MW, respectively. Thus firm size favors the more rapid adoption of large-scale generating units in eastern Canada.

Table 30

Generating Capacity of Large Electric Utility Companies in Canada as of 31 December 1973 (MW)[a]

Province	Company	Capacity
Newfoundland	Churchill Falls (Labrador) Corp. Ltd.[b]	3,325
	Newfoundland and Labrador Power Com.	799
	Newfoundland Light and Power Co. Ltd.[b]	141
Prince Edward Island	Maritime Electric Co. Ltd.[b]	111
Nova Scotia	Nova Scotia Power Corporation	1,108
New Brunswick	New Brunswick Electric Power Commission	1,164
Quebec	Commission Hydroelectrique de Quebec	11,148
Ontario	Hydro-Electric Power Commission of Ontario	16,397
Manitoba	Manitoba Hydro	2,439
Saskatchewan	Saskatchewan Power Corp.	1,612
Alberta	Calgary Power Ltd.[b]	1,899
	City of Edmonton-Edmonton Power	735
	Alberta Power Ltd.[b,c]	514
British Columbia	British Columbia Hydro and Power Authority	4,439

[a] The companies listed for each province account for at least 90% of the generating capacity of privately and publicly owned utilities in the province.

[b] These are privately owned utilities. All other utility companies listed are publicly owned.

[c] Includes 7 MW in the Northwest Territories. The publicly owned Northern Canada Power Commission has 144 MW of generating capacity in the Northwest Territories and the Yukon, accounting for 90% of the total utility generating capacity in those regions.

Source: Statistics Canada, 1974c.

The weighted average firm sizes for the eastern and western regions of Canada are not exactly comparable to those calculated for the United States because small Canadian utilities have not been included in the Canadian averages. Since weighted averages are used, however, the inclusion of the smaller utilities would not greatly reduce the figures for Canada. The two large Canadian utilities, Ontario Hydro and Quebec Hydro, are similar in size to the largest U.S. utilities in 1972—Tennessee Valley Authority (17,714 MW), Commonwealth Edison (13,811 MW), Southern California Edison (11,667 MW), and Pacific Gas and Electric (10,473 MW). The British Columbia Hydro and Power Authority system is about the same size as Consumers Power Company in Michigan (about 4400 MW), but its service area is many times that of the Michigan firm.

Reluctance to have more than 15% of total capacity in any single unit limits the size of new units that most Canadian utilities are willing to install. Only Ontario Hydro and Quebec Hydro are large enough to buy the largest generating units by this criterion. Nevertheless, these two utilities, as noted, account for 57% of Canadian generating capacity.

Since the average cost of fossil fuels and weighted average capacity of utility systems are both larger in eastern Canada than in western Canada, the utilities in eastern Canada should be the more rapid adopters of large-scale generating equipment according to the analysis of Chapter 3.

Comparing regions in Canada with those in the United States, one finds that the eastern region of Canada has slightly lower fossil fuel costs but slightly higher average firm sizes than the Pacific and Middle Atlantic regions. As a result, the adoption patterns found in these three regions should be similar. Western Canadian utilities, which operate in a region of low fuel costs and small system sizes, should compare with utility systems in the Mountain region, which borders some of Canada's western provinces.

Adoption Behavior

Information on the rated capacity, inlet steam pressure, and inlet steam temperature of fossil-fueled steam-electric generating units greater than 10 MW installed by publicly and privately owned utility companies in Canada from 1949 to 1973 was collected from *Electric Power Statistics*, Vol. III, published annually by Statistics Canada. Capacity additions by industrial nonutility companies were not considered. The data for 150

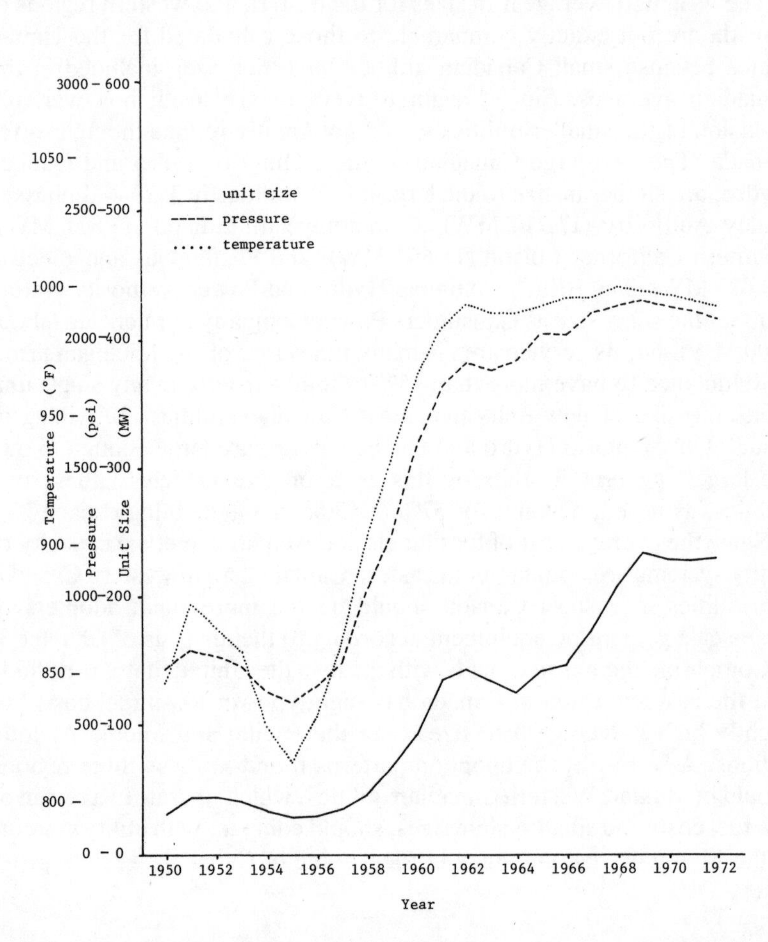

Figure 7. Three-year averages of nameplate capacity, inlet steam pressure, and inlet steam temperature for fossil-fueled steam-electric generating units introduced in Canada, 1950–72.

units installed in Canada over the 25-year period were tabulated by the province in which they were installed. The Northwest Territories and the Yukon were not included in the analysis because together they accounted for less than 1 MW of steam-electric generating capacity at the end of 1973. The correlations between the rated size, pressure, and temperature of the 150 units are high, ranging from .58 to .86 (Table 7).

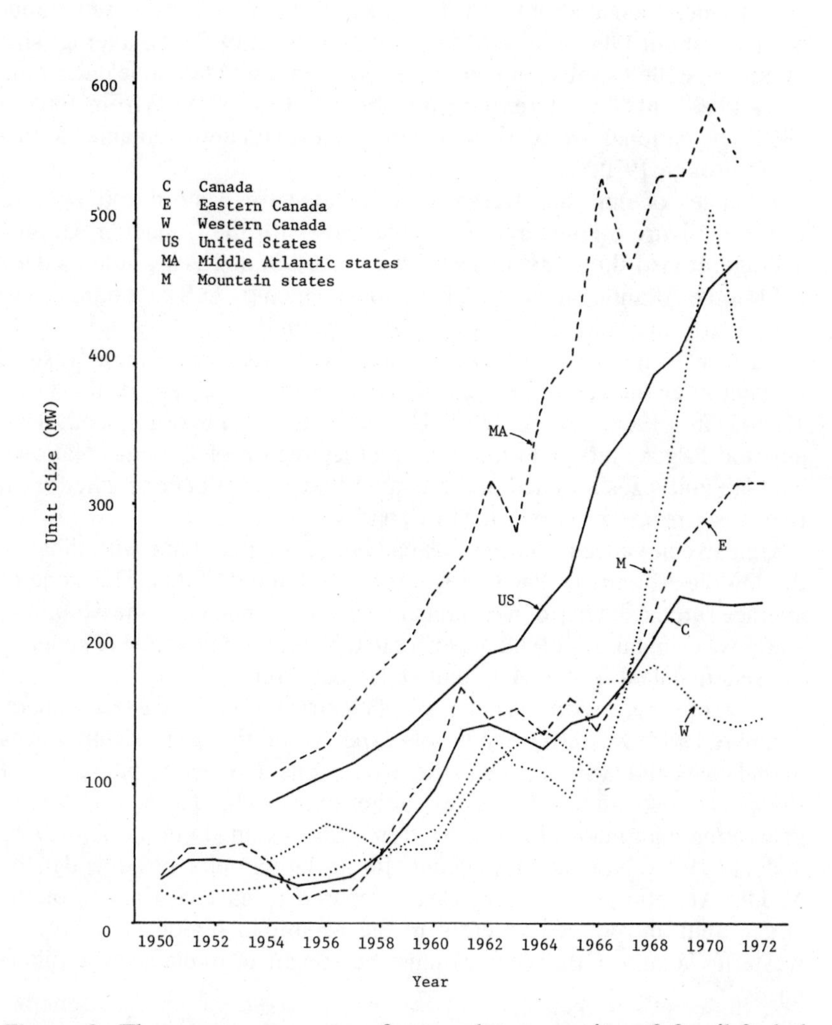

Figure 8. Three-year averages of nameplate capacity of fossil-fueled steam-electric generating units introduced in selected regions of Canada and the United States, 1950–72.

Averages of the characteristics for three-year periods were calculated in the same manner as for units in the United States in Chapter 4, and they are presented in Figure 7. The trend of all three variables over the period under study is roughly similar; they declined from 1951 through 1955 and then increased markedly until 1962. Temperatures have remained fairly steady since then at about 1000°F while pressures have increased gradually from about 1900 psi in 1962 to about 2100 psi in 1972. The average size of units installed showed no increase and even a slight decline during the early 1960s but then increased from about 143 to 237 MW from 1965 to 1969. The national average size of new fossil steam units remained at this level through 1972.

Averages of unit size were also calculated for eastern and western Canada. Figure 8 compares these averages with the Canadian national average. It also shows averages for the United States as a whole, and for the Middle Atlantic and Mountain regions, which by U.S. standards were particularly fast and slow to introduce large units.

Eastern Canada has on average installed larger units than western Canada over the entire period studied with the exception of the years 1955–57 and 1966. Utilities in the United States, on the other hand, have adopted larger units than utilities in either region of Canada. National average unit sizes attained in the United States have been reached from two to six or more years later in Canada.

Unit size advances in eastern Canada have not kept pace with those in the Middle Atlantic or Pacific regions of the United States. The trend of average rated capacity of new units in western Canada and the Mountain states was similar until 1968, when a rapid increase in the scale of generating sets installed in the Mountain states occurred.

The more rapid adoption behavior of electric utilities in eastern Canada compared with western Canada was expected on the basis of differences in fuel costs and system sizes in the two regions. However, differences of fossil fuel costs and system sizes cannot explain the decrease in size of generating equipment that occurred in eastern Canada in the mid-1950s and mid-1960s. Nor can they account for the larger units introduced in the Middle Atlantic and Pacific states compared to eastern Canada or the larger units introduced recently in the Mountain region compared to western Canada. Other factors must be sought to explain these differences.

Other Factors Affecting Diffusion

Types of Generation

Table 31 shows the distribution of hydroelectric generating capacity in Canada. It also indicates that this method of generation still accounts for

Table 31

Distribution of Hydroelectric Generating Capacity of Electric Utilities in Canada, by Province and Region, 1973 (capacity in MW)

Region/Province	Hydroelectric	Total	Percentage Hydroelectric
East	22,987.8	36,240.8	63.4
Newfoundland	4,235.7	4,659.8	90.9
Prince Edward Island	0.0	118.2	0.0
Nova Scotia	155.3	1,112.9	14.0
New Brunswick	665.7	1,207.5	55.1
Quebec	11,176.6	12,127.4	92.2
Ontario	6,754.5	17,015.0	39.7
West	6,817.1	12,139.7	56.2
Manitoba	2,169.1	2,657.3	81.6
Saskatchewan	554.6	1,730.1	32.1
Alberta	718.3	3,249.9	22.1
British Columbia	3,375.1	4,502.4	75.0
North	58.1	160.3	36.2
Northwest Territories	32.0	105.2	30.4
Yukon	26.1	55.1	47.4
Total	29,863.0	48,540.8	61.5

Source: Statistics Canada, 1974c, p. 15.

over 60% of the country's capacity even though its importance has been declining over the postwar period.

Canadian systems with substantial hydroelectric capacity rely on it for base-load operation and use fossil-fueled units on a shift basis (intermediate load) due to the higher operating costs of thermal generation. Fossil steam units so used must be capable of handling a wide range of loads. Part-time variable generation (cycling) with frequent stops and starts is not consistent with the design and construction of large high-pressure, high-temperature, fossil-fueled units. These units are built for base-load operations. When run on an intermittent basis they realize far smaller savings in costs and have relatively high forced outage rates (Table 32 and Table 6).

Thus after taking account of differences in system size and fuel costs, one would expect the fossil-fueled units installed in Canada to be smaller than those in the United States, as was found to be the case. Since the percentage of total capacity accounted for by hydroelectric facilities is

Table 32
Forced Outage Rates of Ontario Hydro Fossil-Fueled Steam-Electric Generating Units

Unit Size (MW)	Number of Units	Average FOR (1965–72)[a]	Equivalent FOR for U.S. Units[b]
66	4	5.4	1.96
100	5	9.4	3.82
200	4	9.2	6.52
300	8	17.0	6.52
500	4	4.4	11.33

[a] Exclusive of effect of 1972 Ontario Hydro strike.
[b] Edison Electric Institute data presented in Table 6 for units in scale class including sizes listed.
Source: G.F. McIntyre, Ontario Hydro, written commun., 1974.

only slightly greater in eastern Canada than western Canada (Table 31), this factor is unlikely to effect differences in the average size of units introduced in these two regions. However, if a finer regional breakdown were employed, Ontario, Saskatchewan, Alberta, and other provinces that rely less heavily on hydroelectric power might be found to install larger fossil-fueled units than Quebec, Newfoundland, British Columbia, and Manitoba.

Ontario Hydro's large share of Canada's fossil-fueled generating capacity, readily apparent from Table 31, helps explain the decrease in the average size of fossil-fueled steam-electric units noted earlier (Figure 7) that occurred during the early 1950s. The Ontario utility added 664 MW of fossil-fueled capacity from 1951 to 1953 and 3300 MW of fossil-fueled capacity from 1959 to 1968. Between 1953 and 1959, however, it added no new fossil-fueled capacity, concentrating instead on the development of new hydroelectric facilities (1400 MW at the Sir Adam Beck station and 912 MW at the Robert H. Saunders station). Because Ontario Hydro is the largest electricity producer in Canada and tends to install large fossil-fueled units, when it fails to introduce such units, the national average unit size for newly introduced units for all Canada falls, as happened in the early 1950s.

Similarly, Ontario Hydro's addition of large nuclear units explains some of the recent leveling off of the average capacity of units installed in Canada (Figure 7). The portion of Ontario Hydro's production accounted for by nuclear power jumped from 1.6% in 1970 to 19.3% in 1973 (AECL Annual Report, 1974, p. 7).

Long-range forecasts for all of Canada include a large amount of nuclear generation. The Canadian government promotes the use of nuclear power through the activities of Atomic Energy of Canada, Limited (AECL), a Crown corporation incorporated in 1952 to promote peaceful uses of atomic energy by subsidizing research and development and by providing financial assistance for construction. This support may have an important future effect on the country's use of nuclear power for base-load generation. If this is the case, utilities are likely to install smaller fossil-fueled units than they would otherwise, and use them primarily for intermediate load or cycling operations.

The entrance of the federal government into the nuclear power program was prompted in part by a desire to exploit domestic uranium resources. Through AECL, the government is promoting the design and construction of nuclear power stations in Canada and abroad, a function performed by the private sector (equipment manufacturers and consulting engineers) in the United States. Canadian nuclear reactors have been installed in developing countries such as Argentina, India, and South Korea (AECL Annual Report, 1974, pp. 9–10).

Financial support for nuclear development is offered in the form of federal government loans to Canadian provinces covering 50% of the capital cost of a prototype nuclear plant. The interest rate on these loans is about 1% less than the provinces could obtain from other sources.

In areas where electricity is produced in large part by hydroelectric or nuclear facilities, the intermediate load role played by fossil steam units appears to constrain the size of generating sets and helps account for some of the differences between Canada and the United States that firm size and fuel costs alone fail to explain. In those Atlantic and Prairie provinces that depend on fossil steam generation for the greater part of their electricity needs, other factors must be sought to explain the installation of small units.

Spatial Distribution of Demand

The markets served by Canadian utilities are for the most part less dense than those of utilities in the United States. The population density in the eastern region of Canada is less than half of that of the West North Central region of the United States, which ranked eighth among the nine U.S. regions. The Mountain region, which is the most sparsely populated area in the coterminous United States, has a population density more than 75% greater than that of the four western Canadian provinces. The remoteness of markets in Canada is accentuated by natural barriers in some areas. The isolated nature of Prince Edward Island points out the problem of using regional population densities to estimate the nature of electricity markets because the difficulty of interconnection with dense markets is not expressed by this variable.

Since the market for electric power in Canada is dispersed, transmission costs are a major consideration in capacity addition decisions. Indeed, the importance of transmission costs in Canada is apparent from the existence of extremely small (less than 1 MW) generating plants in some northern and prairie towns and from Canadian leadership in certain aspects of transmission technology such as long-distance transmission. The reduction of costs of long-distance transmission made possible the tapping of hydroelectric potential in northern Canada including the Peace River (British Columbia), the Nelson River (Manitoba), and Churchill Falls (Newfoundland).

Differences in population density and in the intensity of electric power demand help explain why the United States has tended to introduce considerably larger generating units than Canada even after taking account of differences between the two countries in the size of utilities and fuel costs. In sparsely populated areas the extra cost of long-distance transmission at some point outweighs the saving in generation costs derived from larger generating units, and thus discourages the latters' introduction. Although many of the large Canadian provinces are fairly densely populated along their southern borders, the remote nature of certain markets served by many Canadian utilities, especially those in the Prairie and Atlantic provinces, makes it more economical to install smaller generating units.

Nature of Ownership

In Canada public ownership is the rule, rather than the exception as in the United States. Table 33 shows the percentage of electric utility generating capacity owned by public and private interests for various regions and provinces in Canada. In many provinces public ownership is an outgrowth of provincial responsibility for the development of natural resources, including water power. In Alberta and Prince Edward Island, where utilities were, and still are, primarily dependent on fossil fuels, the industry is mainly investor-owned. Some investor-owned firms have reverted to public control as discussed in Chapter 2. Two Alberta firms—Calgary

Table 33

Private and Public Ownership of Electric Utility Generating Capacity in Canada as a Percentage of Total Capacity by Province and Region, 1973

Region/Province	Public	Private
East	86.2	13.8
Newfoundland	17.7	82.3
Prince Edward Island	5.8	94.2
Nova Scotia	100.0	0.0
New Brunswick	97.4	2.6
Quebec	94.4	5.6
Ontario	98.0	2.0
West	78.8	21.2
Manitoba	100.0	0.0
Saskatchewan	93.8	6.2
Alberta	25.8	74.2
British Columbia	98.9	1.1
North	90.8	9.2
Northwest Territories	93.7	6.3
Yukon	85.2	14.8
Total	84.4	15.6

Source: Statistics Canada, 1974c, p. 15.

Power, Ltd., and Alberta Power, Ltd.—account for 95% of privately owned fossil-fueled generating capacity in Canada (Statistics Canada, 1974c, pp. 15, 77, 78).

The effect of interregional differences in ownership on the size of generating units adopted, if there is any, is difficult to discern. Western Canada, with somewhat higher private ownership than the eastern part of the country (Table 33), does introduce smaller units, but this behavior can be accounted for readily by other factors, particularly lower fuel costs and smaller firm sizes. Nor does the influence of ownership become clearer if one looks at individual provinces. The three provinces where investor-owned utilities control the major share of capacity all tend to introduce small generating units, but again this can easily be explained by other considerations. Prince Edward Island has very small system size, Newfoundland relies heavily on hydroelectric power, and Alberta has widely dispersed demand.

Interest Rates

The increasing role of provincial authorities in all aspects of electric utility operation in Canada makes it difficult to determine whether the cost of obtaining funds for new generation projects varies within Canada. Capital costs for provincially owned utilities may depend on funding from the provincial government, from external markets, and, in some cases, from the federal government. However, neither the Canadian officials interviewed for this study nor other sources consulted indicated that regional differences in interest rates constitute a significant determinant of regional differences in the average unit size of generating units installed in Canada.

Interconnections and Joint Ownership Agreements

Interconnection agreements do not play as great a role in reducing reserve requirements and increasing the scale of capacity additions in Canada as in the United States, largely because the transmission facilities needed to interconnect systems are far less developed in Canada. There is also much lighter population density, and some areas, such as Newfoundland and Prince Edward Island, are completely isolated from other systems. Interconnections between and even within Canadian systems are often not as complete as those in the United States. The Ontario Hydro system is comprised of an eastern subsystem of just under 16,000 MW and a western subsystem of 600–700 MW. The interconnection between the two subsystems is capable of carrying 200 MW of capacity, which is very significant to the reserve picture of the western subsystem but does not substantially alter the reliability situation in the eastern subsystem (J.W. James and G.F. McIntyre, Ontario Hydro, oral commun., 1974).

The Department of Energy, Mines, and Resources encourages interconnections between systems in order to increase system reliabilities and to decrease problems of security of fuel supplies. It is trying to establish a complete national interconnection of transmission systems to the extent possible given technological and economic constraints. After the Department determines what interconnections should be constructed, it recommends federal government financial support that takes the form of a loan covering 50% of their capital cost. The Canadian government also encourages ties between systems in Canada and the United States when utilities in both countries find international transmission links to be of mutual benefit. Despite such efforts, however, interconnections are not extensive in Canada and appear to have done little to alleviate the constraining effect of system size on size of new generating units.

Joint ownership agreements have not been used in Canada to gain additional scale economies since the smaller utilities that could benefit from this type of agreement are in most cases too distant from neighboring utilities to be able to locate a jointly owned facility where the output could be shared to the benefit of all owners. So joint ownership, like interconnections, has had little effect on interregional adoption behavior in Canada.

Growth of Demand

Average annual growth rates of net generating capability[2] in Canada, which reflect the growth in demand for electricity, jumped from 6.6% during 1963–68 to 9.3% during 1968–73. In the United States capacity increased slightly faster during the earlier period and somewhat slower during the later period (Table 27). The difference between the average size of generating units introduced in the United States and Canada, however, has tended to widen over the last ten years, which is just the opposite of what one would expect if the growth of electricity demand were a major factor in the selection of unit size.

Within Canada, generating capability grew 5.7% per year during 1963–68 in eastern Canada and 8.8% in western Canada. During 1968–73, the figures were 9.8 and 8.0%, respectively. Thus if demand growth significantly affects the size of units introduced, one would expect to find a tendency for the disparity in average unit size between eastern and western Canada to widen over the past decade, which in fact has occurred (Figure 8). Thus in contrast with the United States, there is some evidence to suggest that growth in demand may affect the selection of generating units in Canada, though as noted the evidence is conflicting and far from conclusive.

Access to World Equipment Markets

The federal government promotes the domestic production of heavy electrical generating equipment in Canada through the Department of Industry, Trade, and Commerce. This department was established in 1969 to help the Canadian industrial and business community take advantage of new scientific and technological advances so that they could develop products and processes, and then sell them in domestic and foreign markets. A Department of Industry, Trade, and Commerce decision to provide financial assistance to a Canadian firm for research and development on new types of generating equipment can be an important factor in determining if that equipment will be produced domestically and what its price will be relative to units of foreign manufacture. The Department of Industry, Trade, and Commerce helped establish the Howden-Parsons plant in Canada, and government policy seeks to maximize the amount of domestic input in all sectors of the economy. The provincial government in Quebec tries to maximize local input by buying equipment from the Alsthom group (France) and assembling the equipment in Quebec (A.J. Surrey, oral commun., 1974).

The largest thermal units recently introduced in Canada are domestic products. Canadian General Electric constructed the 574-MW oil-fired units installed at Ontario Hydro's Lennox station in 1974 and 1975, and Howden-Parsons produced the first of the 800-MW nuclear steam units for Ontario Hydro's Bruce station (Canadian Department of Energy, Mines, and Resources, 1974, pp. 77–78).

Nevertheless, Canadian utilities are free to choose the size of their generating units on economic grounds and are not constrained to unit sizes offered by Canadian General Electric Ltd. and Howden-Parsons Ltd., two domestic producers of steam-electric generating units. Publicly owned utilities may have to demonstrate that a planned addition is not available or is more expensive from domestic sources before making a foreign purchase, but this requirement does not prevent imports. Turbine-generator sets ranging in size from 150 to 500 MW produced by Hitachi (Japan), C.A. Parsons (Great Britain), English Electric (Great Britain), and General Electric (United States) have recently been installed or will soon be installed in Canadian utility systems.

In summary, the major factors influencing the size of generating units adopted in Canada are system size, spatial distribution of demand, and the type of primary energy used to generate electricity. Interregional differences in fossil fuel costs and in the growth rate of demand for electricity may also be of some importance.

Ontario Hydro, a 16,400-MW system with a large amount of fossil-fueled capacity and high-density markets compared to other Canadian utilities, has been the leader in adoption of scale-related advances in fossil-fueled steam-electric generating technology, and as a result the average-sized generating unit introduced in eastern Canada has for most years exceeded that of western Canada. The country has introduced smaller units than the United States, including those regions in the United States with comparable fuel costs and system sizes.

The large percentage of electricity generated by hydroelectric and more recently nuclear power has inhibited the adoption of very large base-load fossil steam units in many parts of Canada. Smaller units have also been introduced in this country because its population is less dense and its markets more dispersed than in the United States. Continued growth in the demand for electricity and increasing progress in transmission technology should facilitate the adoption of larger units in the future.

6
Diffusion in England and Wales

The adoption of electric power generating units by the Central Electricity Generating Board (CEGB), the public enterprise responsible for the generation of electricity in England and Wales, is examined in this chapter. Costs of fossil fuels and the system size of the CEGB are considered in the first section. On the basis of this information, expectations are developed concerning the rate of adoption of advanced generating units in England and Wales compared with the United States and Canada. The second section investigates the actual record of adoption in England and Wales since 1950 and contrasts this record with that anticipated by the previous section. Comparisons are made with the speed of adoption in the United States and Canada. The influence of other factors on the introduction of scale-related technical advances in fossil-steam generating sets in England and Wales is discussed in the final section.

Fuel Costs and System Size

Most of CEGB's fossil-fueled plants burn coal obtained from the National Coal Board (NCB), another publicly owned enterprise established in 1947. For example, during the 1973 calendar year, the CEGB obtained 99.5% of its 66.3 million tons of coal from domestic sources; of this, 97.1% was from the NCB (*CEGB Statistical Yearbook, 1973–74*, p. 20).

Costs of fuel to the CEGB are presented in Table 34. On a pence per gigajoule basis these costs have risen at an average annual rate of 11.77% during the four-year period ending in 1973–74, a much higher rate than the 2.07% average annual increase during the eight years ending in March 1970. [A gigajoule (Gj) is a metric measure of energy equivalent to 10^9 J or 9.4787×10^5 Btu.] When expressed in U.S. cents/million Btu, the CEGB's costs of fuel appear to have decreased from 1966–67 to 1967–68. This apparent break in trend is the result of currency devaluation; conse-

Table 34

Average Cost[a] of Fossil Fuel Used by the Central Electricity Generating Board, 1957–73

Year[b]	New Pence/ Gigajoule	Cents/ million Btu[c]
1957–58	16.373	48.11
1958–59	16.140	47.83
1959–60	15.362	45.41
1960–61	15.762	46.62
1961–62	16.643	49.31
1962–63	16.644	49.22
1963–64	16.738	49.40
1964–65	16.871	49.67
1965–66	16.928	50.06
1966–67	18.520	54.52
1967–68	18.975	48.18
1968–69	19.127	48.12
1969–70	19.601	49.64
1970–71	23.174	58.52
1971–72	25.771	69.40
1972–73	26.226	64.95

[a] Excluding handling costs.

[b] Year of operation ending 31 March.

[c] Conversion on the basis of United States-United Kingdom foreign exchange rate from *United Nations Statistical Yearbook*, relevant years.

Source: Central Electricity Generating Board, *CEGB Statistical Yearbook*; Electricity Council, 1973b.

quently, comparisons with fuel costs in Canada and the United States are not as meaningful for years close to the devaluation as for earlier and later years.

Fuel costs are far higher for the CEGB than for electric utilities in the United States and Canada. For the year ending 31 March 1973, the CEGB paid an average of 64.95 U.S. cents/million Btu. This was 63% higher than the average cost of fossil fuels to U.S. utilities in 1972 and 84% higher than the cost in Canada for that year. Even the 55.02 cents/million Btu paid for fossil fuels in 1972 by utilities in the New England region, the highest cost region in the United States, was 15% less than the CEGB fuel cost for 1972–73.

The recent discoveries of oil and gas in the North Sea present the British

Table 35

Generating Capacity Introduced by the Central Electricity Generating Board by Type of Equipment, 1969–78

Year[a]		Conventional Steam		Nuclear	Gas	Total
		Coal-Fired	Oil-Fired		Turbine	
1969–70	(MW)[b]	2,186	350	—	243	2,779
	(%)	78.7	12.6	—	8.7	100.0
1970–71	(MW)[b]	2,055[c]	1,385	—	93	3,533
	(%)	58.2	39.2	—	2.6	100.0
1971–72	(MW)[b]	3,305[d]	1,500	645	70	5,520
	(%)	59.9	27.2	11.7	1.2	100.0
1972–73	(MW)[e]	1,941	694	200	85	2,920
	(%)	66.5	23.8	6.8	2.9	100.0
1973–74	(MW)[e]	1,595[f]	516	—	35	2,146
	(%)	74.3	24.1	—	1.6	100.0
1974–78	(MW)[e,g]	2,980	3,640	5,280	620	12,520
	(%)	23.8	29.1	42.2	4.0	100.0

[a] Year of operation ending 31 March.
[b] Output capacity.
[c] Includes 350 MW coal/oil fired.
[d] Includes 870 MW coal/oil fired.
[e] Installed (nameplate) capacity.
[f] Includes 20 MW coal/natural gas fired.
[g] Under construction and scheduled for completion in the four-year period 1974–78.

Source: Central Electricity Generating Board, *CEGB Statistical Yearbook.*

with an alternative domestic source of energy in the future. Table 35 shows that more oil-fired generating plants than coal-fired facilities are scheduled for completion in the four-year period ending in 1978. The future costs of North Sea oil and gas will depend on technological difficulties encountered in extracting the fuel and on whether the government decides to participate in the development of the North Sea through taxation of privately owned companies or through public ownership (T.A. Boley, The Electricity Council, oral commun., 1974).

Nuclear power is another potentially important source of energy. Although only 7.7% of installed capacity in 1973, it accounts for a far larger share of the new capacity scheduled to come on stream in the future (Table 35).

The rapid development of nuclear power and North Sea gas along with government support for the contracting British coal industry are parts of an announced government policy of "making possible a national supply of energy at the lowest total cost to the community, having regard to a whole range of relevant considerations—economic and social—and to national and regional economic policies" (Electricity Council, 1973a, p. 13). The government withheld approval for the conversion of coal-fired generating stations to cheaper oil until 1970, but by then rising oil prices had reduced potential fuel savings.

The flexibility sought by the CEGB in developing a system capable of utilizing the cheapest available sources of energy has not been attained because the government restricts its freedom in the interest of secure fuel supplies (Electricity Council, 1973a, p. 13). The security argument extends beyond the traditional national defense issue to possible problems of unexpected import price rises (Gordon, 1970, p. 313). In addition, Gordon (1970, p. 65) has noted that the British policy of banning coal imports from the United States while allowing substantial oil imports suggests the desire to preserve coal mining jobs is more important than concern over security. In any case, British policy preventing the CEGB from relying more heavily on lower cost imports in large part accounts for the relatively high fuel costs found in England and Wales even compared to energy-deficient areas such as New England.

The CEGB system has grown from 12,841 MW of installed capacity in 1950 to 62,564 MW in 1974 (Electricity Council, 1973b, p. 1). Throughout this period, system size should not have constrained the size of capacity additions as the capacity of the largest available steam-electric generating unit (see Table 3) amounted to no more than 2.6% of the total installed capacity of the British system.

Because the electricity supply system in Britain is consolidated into one publicly owned system, it is much larger than the largest companies or even holding companies in the United States or Canada. The combined capacity of all the utilities in the Pennsylvania-New Jersey-Maryland Interconnection, for example, was equal to only 57% of the CEGB's capacity in 1972.

Since Britain has experienced higher fossil fuel costs and has possessed a much larger average system size than the United States and Canada for the past 20 years, one would expect that the units introduced in Britain would on average be larger than those installed by United States and Canadian utilities over this period.

Adoption Behavior

Information on the rated capacity, inlet steam pressure, and inlet steam temperature of fossil-fueled steam-electric generating units larger than 10 MW installed in the CEGB system from 1949 to 1973 was collected from *Particulars of Plant in Operation at 31 March 1973*, an internal publication of the CEGB. The installed capacity (as opposed to maximum output capacity) of units was used in the analysis as this was comparable to nameplate capacity of United States and Canadian units. The CEGB considers a unit to be "commissioned" when it has operated at 60% of load for 72 continuous hours (W.C. Edwards, CEGB, oral commun., 1974). The unit may then be commissioned at an interim rating with additions to interim capacity being added to system capacity statistics in subsequent years. In order to be consistent with United States and Canadian practices, the final installed capacity was considered as the unit size.

The correlation between the rated capacity and steam conditions of 425 units installed in the CEGB system for the 25-year period ending in 1973 is shown in Table 7. The standardization of unit design is more common in the British electricity supply system than in the United States and Canada, which explains why these correlation coefficients are higher for Britain. Since pressure and temperature are highly correlated with unit size, the latter is a proxy for technological advances associated with all of these characteristics.

Averages of unit size and steam conditions for three-year periods were calculated for CEGB capacity additions in the same manner as for U.S. units in Chapter 4 and are plotted in Figure 9. This graph shows the rapid increase of three-year average unit size from less than 200 MW in 1963 to

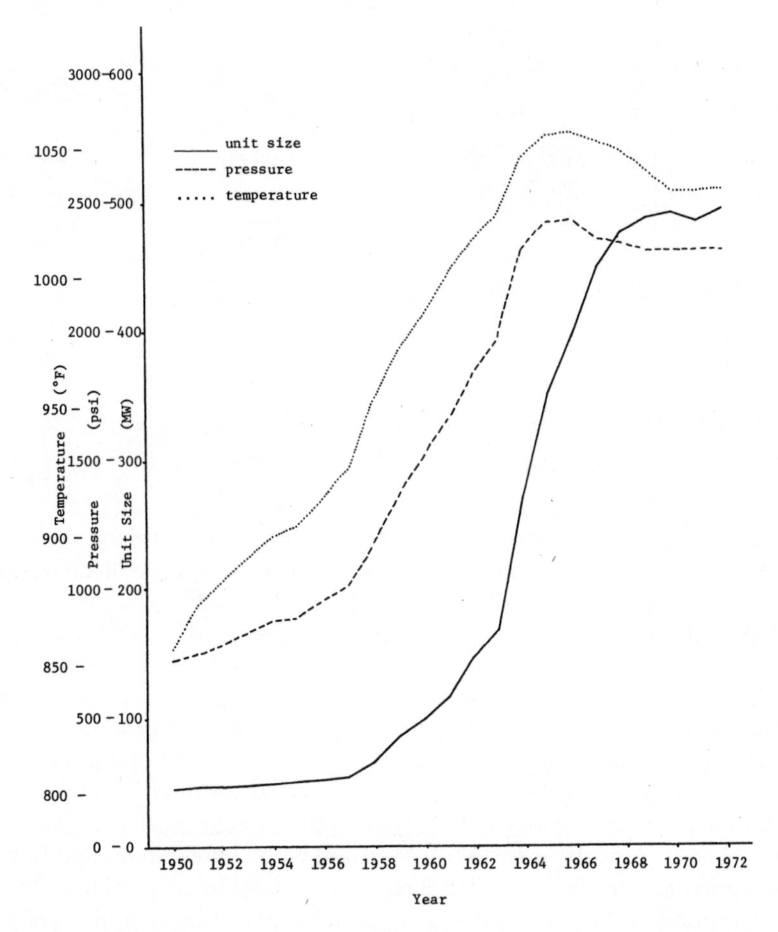

Figure 9. Three-year averages of nameplate capacity, inlet steam pressure, and inlet steam temperature for fossil-fueled steam-electric generating units introduced by the Central Electricity Generating Board, 1950–72.

about 450 MW in 1967. While average unit size has stabilized in recent years, pressures and temperatures have declined slightly since 1965–66. Plans for CEGB expansion through the early 1980s include 660-MW fossil-fueled units featuring subcritical steam conditions (2300 psi, 1050°F).

The average unit sizes installed by the CEGB over the last two decades

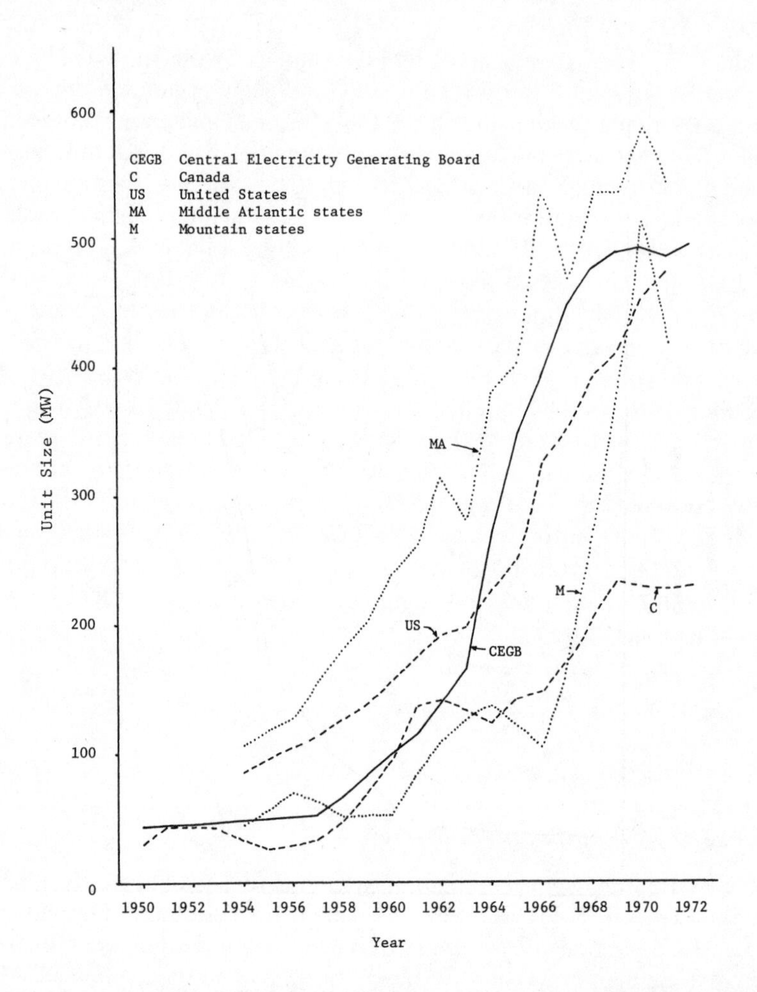

Figure 10. Three-year averages of nameplate capacity of fossil-fueled steam-electric generating units introduced by utilities in Britain, Canada, and the United States, 1950–72.

are compared to similar data for the United States and Canada in Figure 10. The size of fossil steam units installed in Britain was comparable to that in Canada and smaller than that in the United States until about 1963. Following a rapid increase in the mid-1960s, however, the average size of new units in the CEGB system greatly exceeded that of units in Canada and even surpassed the national average for the United States, although certain U.S. regions, such as the Middle Atlantic region (shown in Figure 10), the East South Central region, and the Pacific region, continued to introduce larger units than the CEGB for some or all of the years covered.

The fact that large-scale generating equipment was for a time introduced more slowly in Britain than in the United States, but then advanced more rapidly so that Britain eventually overtook the United States, suggests a corollary is necessary to the hypothesis that large system size stimulates rapid adoption. Specifically, if large system size results in the complete centralization of decision making as is the case in Britain, the system may be slow to abandon old technology, but once the decision is made to do so, it is quick to convert completely to the new technology. The comparison of the United States with the CEGB in Figure 10 suggests that the decision to increase the size of units added to the CEBG system was made after larger units had been adopted by some U.S. utilities. However, once a decision was made to move to a larger scale class, most of the new units introduced were in this class. This does not, though, account for the delay in increasing the size of British generating units prior to 1963. An explanation of this phenomenon requires that other factors affecting diffusion be examined.

Other Factors Affecting Diffusion

Access to World Equipment Markets

The CEGB purchases equipment only from British producers. In 1962 it bought two transformers from a Canadian manufacturer. Despite the successful operation and lower prices of the Canadian equipment, further bids by Canadian firms for CEGB contracts have not been allowed. The CEGB chairman told a Canadian manufacturer in 1968 that reduced requirements for equipment excluded the need for foreign equipment. Required submission of proprietary drawings and exacting manufacturing specifications for CEGB contracts, rather than the more customary performance specifications, present another barrier to the entry of foreign manufacturers into the British market (Epstein, 1972, pp. 15–17).

Exclusive reliance on domestic producers reduces the types of generating units available to the CEGB. Given its size and the importance of its purchases, domestic equipment manufacturers try to respond to its needs, and have even agreed under CEGB persuasion to share contracts, merge, and standardize equipment designs. The CEGB has even forced certain companies to abandon the production of particular types of equipment by failing to award them contracts (Epstein, 1972, pp. 14–15).

Despite this ability to influence domestic equipment producers, complete reliance on them has created problems for the CEGB and impeded the adoption of larger generating units in Britain. This has occurred in particular at those times when the CEGB has decided to increase the size of its standard unit and simultaneously placed a large number of orders for the bigger units. This has forced the domestic producers to build the new units in significant numbers before the technology has been adequately perfected. The following quotes from two 1969 issues of the London *Economist*, quoted in Sporn (1971, pp. 52–53), give an indication of the severity of the problems:

> The power stations of England and Wales have a capacity of 45,000 megawatts, but on last Friday, as the snow came down, a "slight voltage reduction" of 3 percent was ordered by the Central Electricity Generating Board because only 34,500 MW of actual power—77 percent of the total—was available. It is true that November 28th was the trough of a premature cold spell which hit the southern half of Britain some three months earlier than usual—the third week of February was the last cold weather low. But this load shedding was a reminder of just how narrow a power margin is available to the 48.6 million people who depend on the CEGB for their electricity. . . .
>
> Because of what its engineers describe as basic and catastrophic faults in its latest 500 megawatt turbo-generators, the CEGB may have to move from the voltage reductions first described in *The Economist* last week to straight power cuts, with consumers being switched off at source, if the temperature drops to a point where demand swamps the narrow margin of peak-load generating capacity. The reason is simple: of eight turbo-generators each capable, in theory, of delivery 500 MW of power, only one is at present operating. . . .
>
> The CEGB got itself into this mess because it has been scaling up the size of its sets at a prodigious rate in order to get the economies of scale the big sets bring. New batches of designs were being ordered before an earlier generation had been installed, let alone run in. . . . But in 1950 came the decision to order 100 MW and 120 MW sets. This scaling up did not, however, represent any particular technical advance.
>
> That came in 1953–54 when the CEGB decided to buy nine generators each capable of 200 MW. Here two technical leaps were made—which may in the long run prove at the bottom of this winter's grief. One was direct cooling of the copper conductors which act as the link between the revolving stator and the power supply chain, and the other, the increase of boiler operating pressures from 1,500 lb a square inch to 2,300 lb. From there on each increase in the size of generator was decided before the preceding batch of generators had been tried in service. Even now while the CEGB is wrestling night and day to sort out its 500 MW sets, the first of nine new 660 MW units are being installed. Yet, these

units of, for Britain, unprecedented scale were decided on in 1965, two years before the first of the 500 MW sets were commissioned.

The large advance in unit size presented many technical problems to British equipment manufacturers, and by excluding foreign producers, the CEGB failed to benefit from their experience with larger units. In addition, scheduling problems in commissioning new units might not have been as severe had a smaller number of orders for new sets been submitted to domestic manufacturers. A report to Parliament concerning such delays said the large increase in orders for plants of a new design "overwhelmed the technical, production, and managerial resources of the manufacturers and put excessive strain on the staff of the CEGB" (Wilson, 1969, p. 36). Instead of straining domestic manufacturing capacity, orders might have been distributed among domestic and overseas manufacturing firms.

In addition, the reliability problems encountered with the 500-MW units have caused some of the CEGB executives to become disenchanted with purported economies of scale of 1300-MW units. Personal feelings of utility executives can impinge on the decision whether or not to increase unit size and so may retard the adoption of large units in the future (D. Harcombe and J. Syrett, CEGB, oral commun., 1974).

Purchasing policies followed by the CEGB have also had adverse consequences for the domestic equipment manufacturers, who experienced a cutback in demand in the late 1960s due to the previous overordering by the CEGB discussed earlier. Subsequent attempts by the government to orient the electrical manufacturing industry toward export markets came at a time when Britain's share of world trade in electric power equipment was decreasing more rapidly than in other manufactured goods (Surrey and Chesshire, 1972, p. 5). Problems encountered with the initial 500-MW designs installed in the British system and the failure of a British (C.A. Parsons) 500-MW set installed in the Tennessee Valley Authority system of the United States to meet guaranteed performance specification tests did little to spur new orders for British equipment. As of 1973, the 1959 TVA order was the most recent U.S. turbine-generator contract that Parsons had received (Epstein, 1972, p. 50).

Other Types of Generation

If base load is supplied by nuclear or hydro power, fossil steam units assume an intermediate load role in power supply, which reduces the benefits derived from large unit size. This, for example, was found to be the case in Canada and certain regions of the United States. In the past, the CEGB system has relied almost entirely on fossil-fueled generation, so this factor has not discouraged the use of larger generating units.

This may change in the future, however, for as mentioned, the CEGB is attempting to turn increasingly to nuclear power. Since fossil steam units in the 500-MW size range are most consistent with intermediate load generation involving part-time operation and varying loads than are larger fossil steam units, the increasing use of nuclear power for base-load generation is likely to constrain increases in average unit size in the future.

Interconnections, Joint Ownership Agreements, and Growth Rate of Demand

The CEGB system is interconnected with the South of Scotland Electricity Board (SSEB), which had 5399 MW of output capacity in 1972. The SSEB, in turn, is linked with the North of Scotland Hydro-Electric Board, which had 3900 MW of output capacity in 1972. A submarine cable connection was completed across the English Channel to Électricité de France (EdF, the French state-owned electricity supply system) in 1961. This connection allows 160-MW power exchange between EdF and CEGB (Electricity Council, 1973a, p. 16).

There is, however, no reason to believe that these interconnections have increased the size of units adopted in Britain since the largest available units have never amounted to more than 2.6% of CEGB capacity. For the same reason, the absence of joint ownership agreements and the slower growth in electricity demand (as reflected by growth in generating capacity) compared to the United States and Canada have not affected the size of generating units in Britain. Average annual CEGB capacity growth rates were 5.6% for 1963–68 and 6.3% for 1968–73, which compare to 6.6 and 9.3% for Canadian utilities and 6.7 and 8.5% for U.S. utilities over the same periods.

Interest Rates

Capital investment by the CEGB is controlled by the government, which must approve utility expansion plans. In addition to funding from government bonds, surpluses from operations of nationalized industries are reinvested in the industry for self-development and some capital is raised on the European capital markets (M.G. Dwek, CEGB, oral commun., 1974). Increased utility rates are also used as a method of financing expansion plans. Rates are proposed by the CEGB, discussed with the Electricity Council, and published with the approval of the Secretary of State for Energy (formerly the Minister of Power). National financial policies may also affect rates, as when price controls were imposed in recent years.

There is little to suggest that differences in capital costs cause the CEGB to adopt larger or smaller generating units than utilities in North America. The CEGB, like Canadian and United States utilities, fills some of its capital needs from external markets. Given the relative mobility of capital among industrialized countries, the cost of external capital should not differ appreciably for the CEGB and North American power companies. In addition, the British government requires that the CEGB cost the capital provided from public funds at a test discount rate, which is an adjusted private market discount rate (J.P. Hatfield, CEGB, oral commun., 1974).

Nature of Ownership

The record of scale increases in CEGB fossil steam capacity additions shown in Figure 10 suggests that publicly owned electric utilities may be slow to convert to new technological developments but once converted are very swift adopters, though this behavior pattern obviously does not hold for all types of publicly owned utilities. For example, the Tennessee Valley Authority, owned by the U.S. government, has frequently pioneered advances in generating technology, including large-scale fossil-fueled generating sets.

This raises the possibility that it is the centralization of decision making inherent in a single national power system such as the CEGB that primarily influences its adoption behavior rather than the nature of its ownership per se. It is centralization of decision making that permits, and even encourages, the standardization of facilities.

Standardization has long been a major consideration in the selection of unit sizes for CEGB stations. From 1947 to 1950, new steam-electric units in the British system were limited by statute to 30- or 60-MW sets with correspondingly low steam pressures and temperatures. The economies of replication, the high reliability of smaller units, and the time saved by manufacturers in not having to redesign or retool were the economic advantages gained from the policy. Brechling and Surrey (1966) argue that these advantages may have been outweighed by the higher fuel costs entailed over the life of these units compared to larger, more advanced units.

When the utility planners decided that a particular unit best met the national goals of low cost and reliable operation, a "reasonably" large number of these units were installed. This standardization policy held for the 120-MW units that were installed in the late 1950s and early 1960s and was partially in effect for the 200-MW units that were commissioned during the same period. Only a few 350-MW units were installed before the decision was made in October 1959 to install 500-MW units in the CEGB system. The possibility that the British would standardize on these units for an overly long time period was suggested by Brechling and Surrey (1966, p. 41). A combination of events, including the change from 350-MW units without much operating experience and the large number of orders for 500-MW units to British manufacturers (40 were installed from December 1966 to December 1972) resulted in delays in installation and reliability problems with the 500-MW units.

Further size increases have been made in fossil-fueled units installed in the CEGB system. The first of a series of 660-MW units was installed in 1973. The next size increase could occur in the late 1980s when units in the 1200- to 1300-MW range may be added (Electricity Council, 1973a, p. 12).

Thus national ownership coupled with centralized decision making appears to be an important factor explaining the pattern of technological advance in CEGB generating units. Investment decisions have been greatly influenced by a policy of standardization in unit size and by the government's concern over the welfare of the domestic electrical equipment manufacturing industry.

In summary, despite a very large system size and high fuel costs imposed by the use of high-cost domestic coal, the generating units introduced in Britain were on average substantially smaller than in the United States and about the same as in Canada until the mid-1960s. At that time unit size rose sharply, and Britain has since introduced larger units on average than the United States as a whole, though not leading regions in the United States, such as the Middle Atlantic states, nor leading systems, such as the Tennessee Valley Authority and the American Electric Power Company.

There are two principal reasons for the relatively poor performance of Britain. First, decisions regarding new generating capacity are highly centralized in Britain, and this has led to the standardization of units at proven but often relatively low unit sizes. Second, almost exclusive reliance on domestic equipment producers has denied Britain access to the production capacity, expertise, and experience of foreign producers. This has reduced the range of generators available to the country and increased the problems of moving quickly to larger unit sizes.

7
Findings and Policy Implications

In the 1950s, 1960s, and early 1970s diffusion rates of new technology in steam-electric fossil-fueled generating equipment as measured by the average value of unit size and steam characteristics of capacity additions have differed appreciably in the United States, Canada, and England. In none of these countries, though, has the average-sized unit introduced approached the largest established scale class (as defined in Table 3). Average sizes comparable to this scale class have been attained only in the Middle Atlantic, East South Central, and Pacific regions of the United States, which have consistently performed above the national average.

The hypothesis advanced in Chapter 3 that system size and the costs of fossil fuels are the major determinants of unit size predicts that the British Central Electricity Generating Board (CEGB) should be a leader in adopting scale advances since fuel costs are relatively high in Britain and the generating capacity of the state-owned utility dwarfs United States and Canadian firms. This is not the case, however, for the earlier years of the study when the CEGB introduced relatively small generating units. In Britain, as well as in the other two countries, adoption was found to depend not only on system size and fuel costs, but on other factors.

In the United States the size of electric utility companies has been the principal constraint to increases in the size of new units. Coordination between electric utilities through interconnection agreements has offset the inhibiting effect of small system size, but only to a limited extent. Companies have also reduced the constraining effects of system size by joint-ownership agreements. Facilities under joint-ownership include generating units, power stations, and operating companies. However, although jointly owned facilities long predate the 1960s, until 1967 and the construction of several jointly owned facilities by companies in the eastern and southwestern United States (Table 26), their effect was negligible. Even today, their number limits their influence.

Though less important than system size, the cost of fossil fuel is another factor determining the speed of technological diffusion in generating units. The rapid increase in fuel prices since 1972 is likely to heighten concern over the fuel requirements of generating units in the future. Other factors found to influence the size of generating units in the United States include the percentage of hydroelectric or other nonfossil-fueled capacity in an area and, with less certainty, the nature of ownership and population density.

A major factor inhibiting the use of large generating units in many parts of Canada is sparse population density. This condition, coupled with remoteness of service areas, has limited system size and the potential for interconnection agreements in many regions. Advances in transmission technology, however, may relieve this constraint in the future. Also, the growth in demand for electricity service in the Prairie provinces and other lightly settled areas will help.

Another important factor affecting the size unit being installed by Canadian utilities is the nation's commitment to hydroelectric power, and more recently nuclear power. Over 75% of the country's electricity production was generated by hydro and nuclear plants in 1972. In areas where fossil fuels are not used for base-load generation, smaller fossil steam units tend to be installed.

Britain installed large generating sets toward the end of the period studied, but the average size unit installed in the 1950s and early 1960s was far below that of the United States. This behavior is inconsistent with the hypothesis advanced in Chapter 3 that firm size, at least up to a point, and fuel costs are the major determinants of adoption. In Britain, reliance on domestic equipment manufacturers and the complete centralization of decisions concerning the selection of new generating units apparently have outweighed the favorable adoption environment promoted by large system size and high fuel cost. Since the CEGB has always been substantially larger than the minimum size necessary to permit the introduction of the largest available generators, diffusion might have occurred more rapidly had the CEGB allowed four or five regional subdivisions to select independently the new units introduced in their areas.

The preceding discussion indicates that the major reasons why the United States, Canada, and Britain failed to introduce larger units faster than they did differ from one country to another. However, except for Canada with its abundant hydro-power and sparse population density, the major factors impeding adoption are largely institutional and presumably could be eliminated by appropriate government actions.

Moreover, the benefits of taking such actions are great in terms of the fuel and capital that could be saved. Table 36, for example, shows the savings that the United States, Canada, and Great Britain would have realized in 1972 had the national average heat rates for their fossil-fueled

Table 36

National Average Heat Rates and Fossil Fuel Use in the United States, Canada, and Great Britain for Steam-Electric Generation Compared with the American Electric Power System, 1972

Country	Btu ($\times 10^{12}$)	kwh ($\times 10^{9}$)	Btu/kwh	Btu ($\times 10^{12}$) Saved[a]	Savings (%)	Barrels/Day Saved[b]
United States	14,179.45	1,358.78	10,435.4	1,299.5	9.16	605,500
Canada[c]	528.16	48.03	10,996.7	72.9	13.80	33,964
Great Britain (CEGB)	2,098.97	183.26	11,453.5	361.8	17.24	168,599

[a] Savings attainable if all countries had an average annual heat rate equal to that of the American Electric Power Company system (9479 Btu/kwh).

[b] Oil equivalent of Btu savings assuming 140,000 Btu/gal or 5.88 million Btu/barrel.

[c] Data for Canadian fuel use and generation includes internal combustion and gas turbine generation which amounts to 3% of electricity generated by fossil fuels.

steam-electric plants equaled those attained by the American Electric Power Company system. This system has been a world leader in the introduction of large generating units and as a result was the multiple plant system with the lowest heat rate (highest efficiency) in the United States for 1972. The amount of energy that could be saved daily by this improvement in efficiency in the United States is equivalent to over 600,000 barrels of oil, which represents 30% of the 2,000,000 barrel/day ultimate capacity of the Alaskan oil pipeline.

The current concern over the availability of capital for the future expansion of electric power facilities accentuates the need for the rapid adoption of capital-saving technology in this industry. Possible savings in capital appear substantial. The contiguous United States added 36,783 MW of capacity in 1972, of which 16,331 MW were in fossil-fueled steam-electric units. The latter cost over $4800 million (*Electrical World*, 1973, pp. 40, 52). Had these units averaged 815 MW, the largest established scale class at the time, rather than 477 (Table 16), some $260 million in capital expenditures could have been saved (on the assumption that the elasticity of capital costs with respect to unit size is 0.87, as indicated in Table 5). Despite its size, this figure may underestimate the potential capital savings associated with larger unit sizes. Breyer and MacAvoy (1974, pp. 95–96) review some estimates of savings in utility construction costs resulting from the ability to install larger units due to increased coordination among the systems in the United States. These estimates, which included all types of capacity, ranged from $420 million to $930 million/year. This represents from 2.5 to 5.6% of 1972 electric utility capital spending.

The Role of Public Policy

In the United States, the major factor constraining the introduction of large generating units operating at advanced temperatures and pressures is firm size. Except for perhaps ten systems, electric utilities are simply not large enough to install such units without seriously increasing their reliability problems. Public policy can help alleviate this problem in several ways. First, it should encourage interconnection and joint ownership agreements, as the Federal Power Commission has been doing for a number of years through its *National Power Surveys* and other means.

Although the number of such agreements has been increasing in recent years, they have not eliminated the inhibiting effect of small firm size on the adoption of generating units. This suggests public policies that encourage the consolidation of the industry into systems with around 10,000 MW of capacity may be necessary, as Hughes and others have maintained (Hughes, 1971; Breyer and MacAvoy, 1974). This may require a change in the country's antitrust policy, for the Justice Department has intervened in two proposed utility mergers since 1968 and declared a preference for joint ventures over mergers. If such a policy of consolidation were pursued, though, there would be no need to allow the largest ten or twelve utilities to acquire other firms, for they are already of adequate size. (An exception might be made if one of the largest utilities surrounded the service area of a firm of suboptimal size.) And as the British experience demonstrates, excessive consolidation has its drawbacks as well.

Many of the small firms that should merge, however, are either cooperatives or municipals surrounded by privately owned utilities. Merger for them implies a change in the nature of ownership that may be strongly resisted. Here public policy might encourage such small firms to stop generating their own power and to buy it wholesale from neighboring utilities, which are more efficient producers. This should permit the small cooperatives and municipal firms to provide electricity to their customers at lower cost.

In Canada, sparse population density and an abundance of hydropower have slowed the adoption of large generating units. Unlike the situation in the United States, these factors are not institutional in nature, and consequently lie beyond the reach of public policy. This may change in the future with the growth of population and an increase in the relative importance of thermal power generation. At that time, other constraints may become more important, increasing the potential of public policy to foster the adoption of new generating technology.

While the United States has allowed too fragmented a structure to develop in its electric utility industry, England and Wales with a monolithic structure apparently have gone too far in the other direction. The inertial behavior of the CEGB arises in large part from the complete centralization of investment decisions. As already noted, decentralization of such decisions into four or five regions might stimulate earlier adoption of large units. This change might also reduce the large and discrete jumps in orders for generators that have created problems for the domestic equipment producers and the CEGB in the past. Finally, in view of the benefits that could be realized by installing large generating units of foreign manufacture, the British government might consider altering its present restrictions on imports.

Generalizations

The findings and policy recommendations just presented apply specifically to the adoption of large generating units with advanced steam conditions in three countries. This raises the question of whether these findings and recommendations can be generalized to other innovations, industries, and countries.

The marked differences in major determinants affecting diffusion in the three countries examined would caution against attempting to generalize the results for all countries. Although studies of other countries might uncover certain similarities in adoption behavior or even common patterns followed by different groups of countries, the results of this study indicate that no one factor or set of factors universally determines the speed at which new generating units are adopted.

Attempts to extend the results to other innovations and industries seem equally hazardous. Many new innovations in the production of electric power do not necessitate large system size or extensive intersystem coordination and differ in this important sense from large turbine-generator sets. Outdoor production facilities, brushless exciters, stainless steel condenser tubing, and applications of computer-aided statistical techniques to monitor system reliability are examples. Equally, it is difficult to presume that the electric power industry is similar to many other industries given its regulated or state-owned nature, its capital intensity, its absolute reliance on primary energy sources, and the diversity and ubiquity of its customers.

It is always desirable to find conclusions that can be generalized, for results that can be applied across a wide range of innovations, industries, and countries are more useful. Still, when this is not the case, when for an important set of technological advances, such as those associated with large electric power generators, the major determinants of diffusion differ markedly from one country to another, it is important that this be recognized so the public policy can be designed on an *ad hoc* basis and tailored to the specific conditions that exist in each country.

Notes

Chapter 2

1. These units were tandem-compound sets, which have a single generator connected by a shaft to the turbine cylinders. There is one high-pressure cylinder, the size of which increases slowly with the rating of the unit. Low-pressure sections increase in size and number with increasing unit size and an intermediate section may be added (Epstein, 1972, p. 68). Cross-compound units have been used when technical limits such as the ability of the rotor shaft to handle excessive weight prohibit the use of tandem-compound designs. In cross-compound designs, two or more generators are used and each is attached to a separate turbine stage. Unit scale is increased over tandem-compound designs by using some smaller components in a more complex arrangement with cross-compound units (Hughes, 1970, pp. 114–15).

Chapter 3

1. See, for example, Alchian and Kessel (1962), Alchian (1965), Downs (1967), Niskanen (1971), and Shapiro (1973).

2. Electric power companies as listed in the U.S. Federal Power Commission statistical publications rather than holding companies and power pools were used in the regional analysis to avoid as much as possible the extension of power companies across regional boundaries. The influence of holding companies, interconnection agreements, power pools, and other coordination methods on regional adoption behavior is also examined in addition to that of system size of the power companies used by the Federal Power Commission.

Chapter 4

1. Census regions were chosen for the analysis because data could be aggregated by states and then grouped into these regions. This would not be possible with FPC regions because the borders of FPC regions often divide states. The regional treatment of the data also creates problems with power companies that operate in more than one geographic region. The use of the electric companies for which statistics are reported by the Federal Power Commission avoids the problem of regional classification of large holding companies, such as the American Electric Power Company system, which has subsidiaries operating in three geo-

graphic regions. The effect of holding companies on regional adoption behavior is discussed in the section on interconnections.

2. The increase of unit size for generating units installed in the United States corresponds well to a growth equation of the exponential form

$$y(t) = e^{a+bt+u}$$

where $y(t)$ is unit size, the technological change variable which changes over time, e is the Napierian constant, t is the time variable, a and b are parameters of the growth curve, and u is an error term, which is assumed to be random and to have a mean of zero. The log-linear estimating equation has a corrected coefficient of determination $\bar{R}^2$ of 0.992 and the average annual growth rate of unit size,

$$\frac{dy/dt}{y(t)} = \frac{be^{a+bt}}{e^{a+bt}} = b = 10.5\%$$

Chapter 5

1. In Canada, the federal government's decisions involving energy policy are guided by the Energy Development Sector of the Department of Energy, Mines, and Resources. Its function is to develop energy policy on a national basis and to provide inputs into government decisions on the energy supply industry. One of the government agencies that carries out these policies is the National Energy Board (NEB). The NEB administers and controls the movement of energy materials. Exporters of electricity must prove that the energy being sent out of Canada is surplus and that the price received is equal to that which would have to be paid for equivalent generation in the United States.

2. Net generating capability measures the expected power of all available generating equipment of a region during the time of one hour firm peak load for all systems in the region. This may differ from rated capacity of generating facilities for various reasons (Statistics Canada, 1974a, p. 5).

References

Alchian, Armen A. 1965. "The Basis of Some Recent Advances in the Theory of Management of the Firm," *Journal of Industrial Economics*, November, pp. 30–41.

Alchian, Armen A., and Kessel, Reuben A. 1962. "Competition, Monopoly, and the Pursuit of Money," in *Aspects of Labor Economics*, Princeton, N. J.: National Bureau of Economics Research, pp. 157–75.

Atomic Energy of Canada, Limited. 1974. *Annual Report, 1973–74*, Ottawa: Information Canada.

Billinton, Roy, Ringlee, Robert J., and Wood, Allen J. 1973. *Power-System Reliability Calculations*, Cambridge, Mass.: MIT Press.

Brechling, F.P.R., and Surrey, A. J. 1966. "An International Comparison of Production Techniques—The Coal-Fired Electricity Generating Industry," *National Institute Economic Review* (London), no. 36, May, pp. 30–42.

Breyer, Stephen G., and MacAvoy, Paul W. 1974. *Energy Regulations by the Federal Power Commission*, Washington, D.C.: Brookings Institution.

Canadian Department of Energy, Mines, and Resources. 1974. *Electric Power in Canada, 1973*, Energy Development Sector. Ottawa: Information Canada.

Central Electricity Generating Board. 1973a. *CEGB Annual Report 1972–73, Volume 2, Statistical Digest and Detailed Accounts*, Engineering Document Unit, London: Courtenay House.

Central Electricity Generating Board. 1973b. *Particulars of Plant in Operation at 31 March 1973*, Information and Applications Section, October.

Central Electricity Generating Board. *CEGB Statistical Yearbook* (various issues), Press and Publicity Office. London: Sudbury House.

Debanné, J.G. 1974. "Oil and Canadian Policy," in *The Energy Question: An International Failure of Policy*, vol. 2, *North America*, E.W. Erickson and L. Waverman, editors, Toronto: University of Toronto Press, pp. 125–47.

Downs, Anthony. 1967. *Inside Bureaucracy*, Boston: Little, Brown.

Edison Electric Institute. 1973. "Report on Equipment Availability for the Thirteenth-Year Period, 1960–1972," Pub. 73–46, New York, December.

Edison Electric Institute. 1974. "EEI Pocketbook of Electric Utility Industry Statistics," 19th ed. Pub. 73–54, New York.

Electrical World. 1973. "1973 Annual Statistical Report," vol. 179, no. 6, 15 March, pp. 35–66.

Electrical World. 1974. "The Electric Century, 1874–1974," vol. 181, no. 11, 1 June, pp. 43–431.
Electricity Council. 1973a. *Electricity Supply in Great Britain–Organisation and Development*, London: 30 Millbank.
Electricity Council. 1973b. *Handbook of Electricity Supply Statistics, 1973 edition*, Intelligence Section, Secretary's Department, London: 30 Millbank.
Epstein, Barbara. 1972. "Power Plant and Free Trade," in *Realities of Free Trade—Two Industry Studies*, by Duncan Burn and Barbara Epstein, Toronto: University of Toronto Press, pp. 1–128.
General Electric Company. 1974. EUBP Electric Utility Data Bank, computer printout, Schenectady, N.Y.
Goldberger, Arthur S. 1964. *Econometric Theory*, New York: Wiley.
Gordon, Richard L. 1970. *The Evolution of Energy Policy in Western Europe*, New York: Praeger.
Hughes, William R. 1970. "Coordination and Integration in the Electric Power Industry," unpublished monograph.
Hughes, William R. 1971. "Scale Frontiers in Electric Power," in *Technological Change in the Regulated Industries*, William M. Captron, editor. Washington, D.C.: Brookings Institution, pp. 44–85.
International Monetary Fund. 1975. *International Financial Statistics* (monthly).
National Coal Association. *Steam-Electric Plant Factors* (annual), Washington, D.C.: Economics and Statistics Division.
Niskanen, William A. 1971. *Bureaucracy and Representative Government*, Chicago: Aldine-Atherton.
Noll, Roger G. 1971. *Reforming Regulation*, Washington, D.C.: Brookings Institution.
Pennsylvania-New Jersey-Maryland (PJM) Interconnection Agreement. 1974. "Allocation of Forecast Requirements to Parties Hereto," Schedules 2.21, 2.211, 2.212, 2.213, and 2.214, issued 1 April, Norristown, Pa.
Phillips, T.A. 1974. "Effects of Generating Unit Construction Slippage on Adequacy and Reliability of Power Supply," Washington, D.C.: U.S. Federal Power Commission, Bureau of Power Staff Report.
Shapiro, David L. 1973. "Can Public Investment Have a Positive Rate of Return?" *Journal of Political Economy*, March/April, pp. 401–13.
Sporn, Philip. 1968. *Vistas in Electric Power*, 3 vols., Oxford: Pergamon Press.
Sporn, Philip. 1971. *The Social Organization of Electric Power Supply in Modern Societies*, Cambridge: MIT Press.
Statistics Canada. 1974a. *Electric Power Statistics*, Vol. I: *Annual Electric Power Survey of Capability and Load, 1973*, catalogue 57–204 annual, Ottawa: Information Canada.
Statistics Canada. 1974b. *Electric Power Statistics*, Vol. II: *Annual Statistics, 1972*, catalogue 57–202 annual, Ottawa: Information Canada.
Statistics Canada. 1974c. *Electric Power Statistics*, Vol. III: *Inventory of Prime Mover and Electric Generating Equipment as at December 31, 1973*, catalogue 57–206 annual, Ottawa: Information Canada.

Surrey, A.J., and Chesshire, J.H. 1972. *World Market for Electric Power Equipment—Rationalisation and Technical Change*, Falmer, Brighton, Sussex, England: Science Policy Research Unit, University of Sussex.
Tilton, John E. 1973. "The Nature of Firm Ownership and the Adoption of Innovations in the Electric Power Industry," paper presented at the Washington, D.C., meeting of the Public Choice Society, March.
Tilton, John E. 1976. "The Public Role in Energy Research and Development," in *Energy Supply and Government Policy*, Robert J. Kalter and William A. Vogely, editors, Ithaca, N.Y.: Cornell University Press, pp. 99–129.
United Nations. *United Nations Statistical Yearbook* (annual), New York: United Nations Publishing Service.
U.S. Department of Agriculture, Rural Electrification Administration. *Annual Statistical Report, Rural Electric Borrowers* (various issues), REA Bulletin 1–1, Washington, D.C.
U.S. Federal Energy Administration. 1974. *Project Independence Blueprint*, Final Task Force Report—Facilities, Washington, D.C.: U.S. Government Printing Office.
U.S. Federal Power Commission. *Statistics of Privately Owned Electric Utilities in the United States* (annual), Washington, D.C.: U.S. Government Printing Office.
U.S. Federal Power Commission. *Statistics of Publicly Owned Electric Utilities in the United States* (annual), Washington, D.C.: U.S. Government Printing Office.
U.S. Federal Power Commission. *Steam-Electric Plant Construction Cost and Annual Production Expenses* (annual), Washington, D.C.: U.S. Government Printing Office.
U.S. Federal Power Commission. 1964. *National Power Survey*, Washington, D.C.: U.S. Government Printing Office.
U.S. Federal Power Commission. 1971. *The 1970 National Power Survey*, Washington, D.C.: U.S. Government Printing Office.
Wall Street Journal. 1974. "Joint Power Firm Slated by 7 Utilities in New York State," 4 April, p. 16.
Wilson, Sir Alan. 1969. "Report of the Committee on Enquiry into Delays in Commissioning C.E.G.B. Power Stations," presented to Parliament by the Minister of Power by Command of Her Majesty, March, Cmnd. 3960.